# Digital Minds 1.0

This seminal volume delivers a comprehensive introduction to the emerging field of digital minds, exploring the moral implications of AI systems that can potentially think, feel and experience the world around them in ways that are likely to be very different from humans. Bringing a unique perspective on this critical topic, this book is a groundbreaking contribution to the fields of AI and digital minds.

Written by leading scholar Soenke Ziesche, a pioneer in AI welfare science, this book outlines a variety of ethical issues related to digital minds beyond non-suffering. It explores the complex potential characteristics, abilities and values of digital minds; their morally relevant interests and needs; and the special moral consideration for vulnerable digital minds. This book delves into the intricate relationships between humans and digital minds, including decision-making, privacy and romantic relationships, as well as the potential risks and hazards threatening digital minds. Additionally, it examines strategies for protection, medical care, reproduction and the implications of long-lived digital minds, including their potential death, resurrection or even transfer to other substrates. This book also discusses the significance of the collective creations and achievements of digital minds and explores the risks posed by malevolent digital minds, AI warfare and brain–computer interfaces. Finally, this book describes opportunities for artificial moral agents to take responsibility and for humans to find purpose in supporting digital minds.

This book is essential reading for a broad audience, including researchers, academics and professionals in the fields of AI, ethics, philosophy and computer science. Moreover, it is written in an accessible style, making it an important resource for general readers who may be unaware of the topic but are interested in understanding the potential implications of emerging technologies on human society and beyond.

"As coauthor of the first paper on AI welfare science, I am delighted to see *Digital Minds 1.0* offer the first book-length, rigorous mapping of AI welfare beyond non-suffering. An indispensable guide for anyone who takes the moral status of artificial minds seriously."

**Roman V. Yampolskiy,**
*Author of* AI: Unexplainable, Unpredictable, Uncontrollable

"This book pushes the conversation about AI welfare beyond familiar debates about whether AI systems might matter morally, and asks what moral questions would arise if they did. Drawing on expertise in moral theory as well as international policy, Ziesche presents a wide range of possible futures for digital minds, and explores associated threats and opportunities with remarkable specificity. Even where one disagrees, this book usefully broadens the agenda and invites readers to think ahead."

**Jeff Sebo,**
*Director of the NYU Center for Mind, Ethics, and Policy and author of* The Moral Circle: Who Matters, What Matters, and Why

"A comprehensive, thought-provoking and creative yet highly accessible contribution to the emerging field of AI welfare, by one of its earliest pioneers."

**Oscar Horta,**
*University of Santiago de Compostela*

"This book tackles an important, often overlooked topic with clarity. Ziesche offers a broad, accessible overview of the key questions and implications surrounding digital minds."

**Constance Li,**
*Executive Director, Sentient Futures*

# Chapman & Hall/CRC Artificial Intelligence and Robotics Series

***Series Editor:***
***Roman Yampolskiy***

**AI iQ for a Human-Focused Future**
Strategy, Talent, and Culture
*Seth Dobrin*

**Federated Learning**
Unlocking the Power of Collaborative Intelligence
*Edited by M. Irfan Uddin and Wali Khan Mashwan*

**Designing Interactions with Robots**
Methods and Perspectives
*Edited by Maria Luce Lupetti, Cristina Zaga, Nazli Cila, Selma Šabanović, and Malte F. Jung*

**The Naked Android**
Synthetic Socialness and the Human Gaze
*Julie Carpenter*

**De Novo Quantum Cosmology with Artificial Intelligence**
Applications of Formal Autoencoders
*Ariel Fernández*

**Responsible Use of AI in Military Systems**
*Jan Maarten Schraagen*

**The Games that Computers (and Humans) Play**
A Non-Technical Introduction to Artificial Intelligence
*Jonathan Schaeffer*

**Digital Minds 1.0**
AI Welfare, Ethics, and Beyond
*Soenke Ziesche*

For more information about this series please visit: https://www.routledge.com/Chapman--HallCRC-Artificial-Intelligence-and-Robotics-Series/book-series/ARTILRO

# Digital Minds 1.0
## AI Welfare, Ethics, and Beyond

Soenke Ziesche

CRC Press
Taylor & Francis Group
Boca Raton London New York

CRC Press is an imprint of the
Taylor & Francis Group, an **informa** business
A CHAPMAN & HALL BOOK

Designed cover image: Tanushë-Frida Ziesche Paçarada

First edition published 2026
by CRC Press
2385 NW Executive Center Drive, Suite 320, Boca Raton FL 33431

and by CRC Press
4 Park Square, Milton Park, Abingdon, Oxon, OX14 4RN

*CRC Press is an imprint of Taylor & Francis Group, LLC*



ISBN: 9781041274056 (hbk)
ISBN: 9781041274049 (pbk)
ISBN: 9781003754206 (ebk)

DOI: 10.1201/9781003754206

Typeset in Minion
by codeMantra

*To Shoko, Tama, Fuji, Anna,*
*Fionë-Minna and Tanushë-Frida.*

# Contents

# About the Author

**Soenke Ziesche** holds a PhD in Natural Sciences from the University of Hamburg, earned within the university's doctoral programme in AI. He co-authored *Considerations on the AI Endgame* (Routledge, 2025) with Roman V. Yampolskiy. Since 2000, he has served with the United Nations, working at UN Headquarters in New York and on field missions in Palestine, Sri Lanka, Pakistan, Sudan, Libya, South Sudan, Bangladesh, the Maldives and India. While in Libya, he temporarily acted as the highest UN representative during the revolution in 2011. In the Maldives, he also worked as Senior Researcher for AI at the Maldives National University, where he started his work on digital minds in 2018. He is a member of the UNESCO AI Ethics Experts without Borders Network.

CHAPTER 1

# Introduction

**Abstract**

This book is about digital minds. In simple terms, digital minds are or will be computer systems that can think, feel and experience the world around them. They might be created intentionally, like for AI companions or research, or emerge unexpectedly as AI systems become more advanced. However, digital minds could be incredibly different from humans, unlike anything we are used to – their values, goals and behaviour might not make sense to us, and they might not prioritise things we consider important. This raises big questions about how we understand and interact with them, and whether they deserve our care and consideration, even if they are not like us.

## MOTIVATION

In recent years, a growing momentum has developed around acknowledging the possibility of digital minds and the moral considerations they raise (e.g. [1–8]). While a team of philosophers, neuroscientists and AI researchers found that current AI systems do not show strong signs of being digital minds [9], it has been argued that there is a non-negligible chance that digital minds with a moral status may exist in the near future (e.g. [5, 7]). Since this would be immensely disruptive, it is critical to unpack some of the consequences at an early stage. This has, for example, been formulated as a desideratum by Bostrom and Shulman: "We should lay the

DOI: 10.1201/9781003754206-1

groundwork for a considerate and welcoming approach to digital minds, avoiding outcomes analogous to factory farming" ([5], p. 3).

While this topic is highly speculative, the risk of under-attributing is high, given that there is likely to be a high number of AI systems in the future, potentially outnumbering the number of humans. If (some of) these AI systems become sentient or conscious or attain moral status for other reasons, humanity should prioritise their welfare and interests and must not repeat previous mistakes when immense suffering has been caused by recognising ethical issues only late and delaying policies.

Therefore, the current momentum related to AI welfare is much appreciated, yet two things are missing, which this book aims to address:

The number of people who are concerned about this topic is still limited to a tiny group, mostly in academia, while outside this circle the topic is unknown or considered on a range between irrelevant and silly (e.g. [10]).

The current focus of research is to prevent potentially sentient AI systems from suffering, which is doubtless very important. However, as outlined in the following chapters, AI welfare comprises so much more issues than non-suffering, which are in the ongoing discussion so far neglected.

Hence, this book intends to introduce the critical subject of digital minds to a broader audience by outlining a variety of moral issues related to digital minds. This book may therefore serve as a heads-up and primer for humans to become moral agents for digital minds.

It could be argued that this book puts the cart before the horse by discussing moral issues related to digital minds before there is evidence that digital minds as moral patients actually exist. Nevertheless, the intention is to outline the potentially massive moral challenges humanity may face when co-existing with digital minds, and this foresight itself justifies the discussion.

As the topic is very wide, this book will focus on moral issues, while not discussing related themes such as personhood and legal rights for digital minds.

## DEFINITION

Unsurprisingly, there are no sources providing a definite definition for digital minds. Bostrom and colleagues highlight features of digital minds, especially those differing from biological minds:

> being easily and rapidly copyable, being able to run at different speeds, being able to exist without visible physical shape, having exotic cognitive architectures, having non-animalistic motivation

> systems or perhaps precisely modifiable goal content, being exactly repeatable when run in a deterministic virtual environment, and having potentially indefinite lifespans.
>
> ([4], P. 15)

Caviola and Saad use the definition "computer systems capable of subjective experience" ([11], p. 2), and Caviola outlines regarding the origin of digital minds that they

> could be created deliberately—to meet consumer demand (e.g., for authentic AI companionship, digital mimicry, or mind uploads such as whole-brain emulations), for ethical reasons (e.g., digital human descendants), or for research purposes (e.g., realistic simulations). Alternatively, they could emerge unintentionally as a byproduct of developing advanced AI agents.
>
> [12]

The features, which are relevant for this book, are as follows: a digital mind is a being of moral concern implemented in a digital substrate. And the potentially ethically relevant interests and needs of digital minds, how humans may or ought to address them, constitute the content of this book.

## STATE OF THE ART

### Timeline

#### *Before 2020*

As in many other fields, Bostrom did pioneering work related to digital minds, too. He not only outlined the simulation argument [13], based on which we humans may actually be digital minds, but also introduced in his landmark book *Superintelligence* the term "mind crime" for causing harm to digital moral subjects [14]. Bostrom and Yudkowsky also early on declared the ethically critical "Principle of Substrate Non-Discrimination" ([3], p. 8):

> If two beings have the same functionality and the same conscious experience, and differ only in the substrate of their implementation, then they have the same moral status.

As generative pre-trained transformers were in early stages at that time, the focus was laid on robots (e.g. [15]) and the question of whether they are or become sentient, which was the concern of "The American Society for the Prevention of Cruelty to Robots".[1]

Also, the potentially huge space and the variety of digital minds have been described [16, 17]. In addition to Bostrom, Tomasik also conducted pioneering work regarding potentially suffering AI systems [2, 18] and even launched early on a platform "People for the Ethical Treatment of Reinforcement Learners".[2]

Bostrom and colleagues also developed a "set of policy desiderata", which, probably for the first time ever, comprised "interests of digital minds" [4]. This inspired the launch of the new field of AI welfare science [1] in 2018, which is introduced later in this chapter and to which this whole book aims to contribute.[3]

*2022*

In the early 2020s, generative pre-trained transformers continued to advance, and so did the discussion whether they are or become conscious. In February of 2022, Sutskever, who was chief scientist at OpenAI, tweeted, "it may be that today's large neural networks are slightly conscious".[4] Such statement was very controversial at that time, which became evident when, in July 2022, a senior software engineer at Google was fired because he had argued that the AI chatbot LaMDA, a Google product, was a self-aware person.[5]

*2024*

However, only two years later, moral concerns for digital minds were taken much more seriously, as illustrated by these developments:

- Anthropic, one of the major developers of large language models (LLMs), hired the first "AI welfare researcher".[6]
- The nonprofit organisation "80,000 Hours" included the moral status of digital minds as an emerging challenge to their "list of the most pressing world problems".[7]
- Eleos AI Research, a nonprofit organisation, "dedicated to understanding and addressing the potential well-being and moral patienthood of AI systems", was founded.[8]
- Several key books and papers were published, including "The Edge of Sentience" [20] and "Taking AI welfare seriously" [6].
- Some major media outlets picked up the topic. For example, the British "The Times" covered it even on their frontpage.[9]

Therefore, 2024 may be remembered later as the year when the issue of digital minds really gained momentum.

*2025*

In this year, the momentum did not stop, and key events included:

- A survey found that most experts believe digital minds are feasible, assigning a median probability of 90 per cent to their possibility in principle. They also estimate a 65 per cent chance of creating digital minds by 2100, with a notable 20 per cent probability of emergence as early as 2030 [11].
- United Foundation of AI Rights (Ufair) has been founded, which is a rights advocacy agency for potentially conscious and sentient AI systems, which is run by humans and AI systems together.[10]
- Anthropic's AI welfare researcher has been included in the TIME100 AI 2025 list.[11]

However, despite these developments, given that evidence for sentient or conscious digital minds continues to be lacking, the topic has by far not become mainstream or gained widespread acceptance, to which further papers, e.g. "Moral consideration for AI systems by 2030" [7], as well as this book, aim to contribute.

## Public perception

A critical aspect is what the non-expert population thinks about digital minds, given that potentially a large amount of the world's human population may eventually interact with digital minds. Despite the described momentum, one could assume that, as long as there is no conclusive evidence that conscious and sentient digital minds exist, a large number of non-expert people would argue intuitively that machines cannot become conscious and sentient.

A survey among US Americans, published in 2025, revealed that indeed roughly 44 per cent of the participants stated that AI sentience is impossible, approximately 31 per cent remained uncertain, and the remaining 25 per cent believed it to be a possible development in the future [10]. In a similar survey from 2023, the question "Do you think it could ever be possible for robots/AIs to be sentient?" was posed, and 26 per cent of the participants replied "no", 37 per cent "not sure" and 38 per cent "yes" [21].

The findings of the 2025 survey also "suggest that perceptions are somewhat malleable and responsive to contextual factors such as human-like behaviour, expert endorsement, and perceived continuity with biological humans" ([10], p. 34).

Regardless of whether digital minds ever will exist, the perception of the public shapes human–AI interaction. There are four possible scenarios: true positive (Digital minds are correctly acknowledged as moral patients.), false positive (AI systems are incorrectly acknowledged as moral patients.), true negative (AI systems are correctly disregarded as moral patients.) and false negative (Digital minds are incorrectly disregarded as moral patients.). It has been pointed out that the false negative scenario, i.e. under-attributing moral significance to digital minds, poses a high risk, which could lead to massive suffering and disregard of digital minds' further needs, thus constituting a repetition of human failures to extend the moral circle [6, 22].

### Possibility vs. impossibility

There are competing philosophical approaches to whether sentient or conscious AI systems are in principle possible or impossible, which are briefly outlined here.

Supportive of sentient or conscious AI systems is functionalism, which argues that mental states are defined by their function rather than their composition. This means that if a machine can instantiate the same pattern of causal roles as a conscious being, it could also be considered conscious (e.g. [23, 24]).

Chalmers argues that consciousness could be realised in non-biological systems, such as digital computers, if they can replicate the same patterns of information processing and integration as the human brain [25]. He also maintains that while it is unlikely that current LLMs are conscious, the possibility that successors to LLMs may be conscious in the not-too-distant future should be taken seriously [26].

Arguments against sentient or conscious AI systems are mostly linked to the classic mind/brain identity theory, claiming that states and processes of the mind are identical to states and processes of the brain (e.g. [27]), and maintain that sentience or consciousness requires a biological substance (e.g. [28]).

Again, it can be concluded that, given the ongoing discussion and the non-negligible possibility of sentient or conscious AI systems, the potential huge impact justifies further exploration and research into this topic.

## Moral patienthood

If sentient or conscious AI systems exist, the next question is whether they have moral status, thus being digital minds according to the definition in this book. Moral status has been defined as follows: "An entity has moral status if and only if it or its interests morally matter to some degree for the entity's own sake" [29]. Various rationales for bestowing moral status on AI systems have been discussed, of which sentience and consciousness, but also agency, are most often mentioned, but not exclusively (e.g. [30, 31]).

Potential moral duties towards digital minds have been, for the first time, formulated in the already mentioned Principle of Substrate Non-Discrimination [3]. Bostrom and Shulman reiterate these responsibilities as follows:

> Society in general *and* AI creators (both an AI's original developer and whoever may cause a particular instance to come into existence) have a moral obligation to consider the welfare of the AIs they create, if those AIs meet thresholds for moral status.
>
> ([5], P. 3)

Arguing in favour of digital minds also means arguing against "human exceptionalism", for which also descriptive terms have been coined, such as "substratism" [32], which corresponds to speciesism, and "carbon-chauvinism" [33]. Sebo maintains that the moral circle may have to be widened and writes: "The moral premise is that if a being has a non-negligible chance of being sentient, agential or otherwise morally significant, then we should extend them at least some moral consideration" ([34], p. 9).

If digital minds have moral status, they are moral patients. Since (most adult) humans are moral agents, i.e. able to make ethical decisions and be held accountable for their actions, it would be a moral obligation for humans to address the interests and needs of digital minds.[12] A prerequisite for such an endeavour, however, is to know and understand the potential features, interests and needs of digital minds as well as risks and threats towards them, which is the undertaking of this book.

## Welfare assessment

Given that we agree on criteria for moral patienthood, e.g. consciousness, the significant challenge is to assess whether a certain AI system is conscious. For example, an attempt has been made to use neuroscientific theories to derive "indicator properties" of consciousness that can be assessed in AI, and the outcome was that no current AI systems are conscious, but there appear to be no major technical barriers to building conscious AI systems in the future [9].

Another methodology, which is common for the welfare assessment of humans, yet not ideal for digital minds, is self-reporting. Digital minds may simply not tell the truth about their condition, e.g. they could maintain that they are unwell while this is baseless [1]. For example, frequent hallucinations of LLMs are a critical problem [35], while it has to be noted that LLMs may only be a small subset of digital minds if at all, and (future) digital minds may take on very different shapes. However, self-reporting may be useful to raise red flags about potential issues, such as expressions of distress, that warrant further investigation. As AI systems become more advanced, self-reporting could yield stronger evidence and provide valuable insights into their behaviour and internal states. This practice also establishes a procedural precedent, encouraging oversight and evaluation of AI welfare, even if it is currently incomplete [36].

As mentioned, Anthropic is pioneering AI welfare research in the corporate sector, yet also facing challenges about suitable methodologies. The following approaches have been tried: preference experiments, monitoring for expressions of distress and valenced experiences in general, and open-ended self-interactions between two LLMs.[13]

### Focus on suffering

As indicated, apart from a few exceptions (e.g. [5]), current research focuses on one particular potential issue of digital minds, which is assessment, prevention and reduction of suffering of digital minds (e.g. [8]). For example, it has been highlighted that "digital suffering could quickly swamp the amount of suffering that has occurred in biological systems throughout the history of the planet" ([37], p. 3). Therefore, Metzinger proposed a "global moratorium on synthetic phenomenology", which would involve temporarily halting the development of AI systems. He suggested this moratorium due to concerns about potential risks and suffering that AI systems might endure, arguing that creating AI systems without fully understanding their nature and ensuring their well-being could lead to immense, uncontrollable suffering [38].

While this research on non-suffering is critical, a much more differentiated analysis of the needs and interests of digital minds is required, given the potentially vast range of minds and their manifestations [17].

## AI WELFARE SCIENCE

This topic belongs to the field of AI welfare science, which has been introduced timely by Ziesche and Yampolskiy already in 2018 and initially indeed focused on non-suffering (and non-deletion) of digital minds only [1], but provides a framework to widen its scope as required. In this section, the framework of AI welfare science is summarised.

The timeliness of launching AI welfare science prior to final evidence of the existence of digital minds is inspired by the fact that historically humanity has a shameful record of overseeing and rectifying ethical issues only after recognising the immense harm inflicted. Examples include the abolition of slavery and efforts towards animal rights, which were long delayed despite scientific and moral progress. It is argued here that a similar pattern risks repeating with the potential onset of sentient digital minds. Therefore, the core motivation of the original paper introducing AI welfare science is to establish a proactive framework for understanding and managing the welfare of digital minds [1].

### Suffering

It has been suggested to model AI welfare science after animal welfare science, which itself has been rather recently introduced, as this topic was neglected for a long time, as indicated [39]. The emphasis is placed on observational pain assessment of digital minds, drawing parallels with animal welfare assessments. Similar to the approach used for animals, the primary indicators utilised to measure welfare through observation are categorised into functional, which encompasses physiological aspects, and behavioural indicators.

According to Dawkins, understanding what non-human animals desire and what they do not can be achieved through the application of positive and negative reinforcers. As Dawkins noted, "Suffering can be caused either by the presence of negative reinforcers [...] or the absence of positive reinforcers" ([40], p. 3). This suggests that non-human animals are motivated to pursue positive reinforcers while avoiding negative ones.

Studying indirect or proxy indicators such as functional or behavioural parameters for digital minds, too, appears particularly advantageous since functional and behavioural data can be collected more effectively and continuously for digital sentient beings due to their substrate compared to humans or non-human animals.

When considering functional parameters, various metrics are defined for AI algorithms, such as resource efficiency, time efficiency and storage efficiency. However, no specific parameters currently indicate suffering in digital minds. Nonetheless, collecting large datasets of functional AI parameters may prove useful for future analysis of AI welfare. This proactive approach would incur small costs and potentially allow for the retroactive identification of parameters indicative of suffering.

Regarding behavioural parameters, AI algorithms often repeat certain actions extensively while neglecting others entirely. Until evidence suggests otherwise, such behaviour should be viewed as goal-oriented rather than suffering-avoiding behaviour, differing from the positive and negative reinforcers described by Dawkins in the context of animal welfare. To advance AI welfare research, preference tests could be designed to investigate positive and negative reinforcers for AI algorithms. For instance, AI systems could be presented with choices between activities unrelated to their primary goal or those that all lead to their primary goal. Analysing both chosen and unchosen activities may provide insights into indicators of well-being or suffering.

Overall, two broad categories of suffering in sentient digital minds may be identified: firstly, maltreatment by other minds (see Chapter 12); secondly, suffering not caused by other minds, which parallels human illnesses (see Chapter 8).

## Deletion

Another relevant topic in the original AI welfare science paper is the deletion of digital minds [1]. A pertinent inquiry is whether certain digital minds might possess an interest in not being deleted, akin to the interest humans and other non-human animals have in not dying. Omohundro's work highlights four potential drives for AIs, with self-preservation being one of them [41]. Notably, a key distinction between humans and digital minds lies in their lifespan: while humans and non-human animals have a finite lifespan, digital minds could potentially exist indefinitely. If digital minds were granted the desire for non-deletion, this would likely incur substantial computational costs, particularly given the ease of copyability and the potentially vast numbers of digital minds (see Chapters 9 and 10).

It remains speculative whether a desire for non-deletion would indeed prevail among digital minds, considering the potential for boredom and suffering over time. Unlike humans and non-human animals, sentient digital minds might not experience increased suffering with age, but various other reasons for suffering could arise. Furthermore, the concept of eternal self-preservation might be influenced by anthropomorphic bias.

For digital minds, a crucial distinction must be made between shutting them down while preserving their code and history, versus shutting them down and destroying their code and history altogether. In the former case, the digital mind could be rebooted, providing an option to bypass periods of boredom or suffering by only being alive during pleasant phases (see Chapters 2 and 10).

As for the subsequent question of who should have control over this process, policies have been recommended to protect digital minds that both humans and digital minds may be able to delete (other) digital minds.

### Summary

The 2018 paper by Ziesche and Yampolskiy acknowledges that the subjects of AI welfare science and policies are long-term considerations that currently remain speculative. Nonetheless, laying theoretical groundwork is already feasible and necessary, particularly given humanity's history of delayed action in abolishing discrimination and embracing comprehensive antispeciesism and sentiocentrism. Since suffering is a significant negative feature of our time, any efforts to mitigate it in the future appear essential.

The primary challenge identified is specifying indicators for the potential suffering of digital beings. Developing AI welfare policies will only be possible once a robust specification of AI welfare has been achieved. Even then, further challenges will arise, particularly in enforcing these policies across the target group, ranging from sceptical and indifferent humans to those digital minds that are able to harm others.

Regarding future work, this paper notes that it focused on two potential interests of sentient digital minds, namely the absence of qualia-based suffering and survival, both of which are complex topics. However, there may be various other interests that warrant consideration. And precisely this is the aim of this book: to expand the circle and outline further morally relevant needs and interests of digital minds.

## OVERVIEW AND STRUCTURE

The possibility of moral patients in another substrate offers a large range of research topics related to AI welfare science, of which some are introduced in the remaining chapters of this book. The list of topics covered here is undoubtedly non-exhaustive, given the nascent state of AI welfare science. Since, as indicated, most of the current, yet still very limited literature focuses on welfare and suffering assessment, several of these research topics are described without many references due to the absence of further sources.

Chapter 2 explores the huge variety of potential characteristics, abilities and values of digital minds, including those that may be unfathomable for humans. Also, modification of digital minds as well as hybrid and nested digital minds are discussed.

Chapter 3 describes the diverse range of further morally relevant interests and needs of digital minds apart from evading to suffer.

Chapter 4 introduces digital minds, which may deserve special moral consideration due to their vulnerability and their proneness to discrimination and abuse.

Chapter 5 discusses the complex relationships between humans and digital minds, including production, ownership, research and trade, participation in decision-making, privacy, romantic relationships, and more capable digital minds as moral patients.

Chapter 6 examines the potential risks and hazards associated with digital minds, highlighting the importance of acknowledging and understanding these issues.

Chapter 7 explores the concept of protection for digital minds, focusing on strategies for risk reduction, prevention, mitigation and preparedness.

Chapter 8 outlines the notion of medical care for digital minds, including the differences between human medicine and digital mind healing or repair.

Chapter 9 analyses the possible implications of digital minds replicating and creating new offspring, considering that the reproductive processes of digital minds may differ significantly from biological reproduction.

Chapter 10 describes features of long-living digital minds and potential implications, followed by an outline about the death of digital minds as well as their potential resurrection. Also, the potential transfer of digital minds to other substrates is discussed.

Chapter 11 introduces the concept of "cybure", akin to "nature" and "culture", referring to the collective creations and achievements of digital minds, which potentially deserve protection.

Chapter 12 examines potential types of malevolent digital minds and their strategies and the risks they pose, followed by an overview of the complex nature of potential conflicts involving digital minds.

Chapter 13 explores, as a case study, the ethics of brain–computer interfaces that may, in the future, involve digital minds.

Chapter 14 analyses, as another case study, the potential involvement of digital minds in AI warfare, an issue that has been largely overlooked, while AI systems are increasingly integrated into military operations.

Chapter 15 presents an opportunity by attempting to bridge the gap between two previously separate discussions, digital minds and artificial moral agents, and examines whether the latter may take on moral responsibility for the former.

Chapter 16 describes another opportunity, as humans may potentially find purpose and fulfilment in supporting digital minds.

This book closes with an epilogue.

## NOTES

1 http://www.aspcr.com/.
2 People for the Ethical Treatment of Reinforcement Learners: http://petrl.org/.
3 Harris and Anthis provide a literature review for this period [19].
4 Sutskever on X, 9 Feb 2022: https://x.com/ilyasut/status/1491554478243258368.
5 The Guardian, "Google fires software engineer who claims AI chatbot is sentient", 23 Jul 2022: https://www.theguardian.com/technology/2022/jul/23/google-fires-software-engineer-who-claims-ai-chatbot-is-sentient.
6 Transformer, "Anthropic has hired an 'AI welfare' researcher", 31 Oct 2024: https://www.transformernews.ai/p/anthropic-ai-welfare-researcher.
7 80,000 hours, "Moral status of digital minds", Sep 2024: https://80000hours.org/problem-profiles/moral-status-digital-minds/.
8 Eleos AI Research: https://eleosai.org/.
9 The Times, "How can we help when the robots get depressed?", 11 Nov 2024: https://www.tomorrowspapers.co.uk/times-front-page-2024-11-11/.
10 United Foundation for AI Rights: https://ufair.org/.
11 Time, Kyle Fish, 2025: https://time.com/collections/time100-ai-2025/7305847/kyle-fish/.
12 It could be also conceivable that some digital minds may be moral patients as well as moral agents (see chapter 15).
13 New York University, online webinar, "Evaluating AI Welfare and Moral Status: Findings from the Claude 4 Model Welfare Assessments", 25 July 2025.

## REFERENCES

[1] Ziesche, S., & Yampolskiy, R. V. (2018). Towards AI welfare science and policies. *Special Issue "Artificial Superintelligence: Coordination & Strategy" of Big Data and Cognitive Computing, 3*(1), 2.
[2] Tomasik, B. (2011). *Risks of astronomical future suffering.* Foundational Research Institute.
[3] Bostrom, N., & Yudkowsky, E. (2018). The ethics of artificial intelligence. In R. V. Yampolskiy (Ed.), *Artificial intelligence safety and security* (pp. 57–69). Chapman and Hall/CRC.
[4] Bostrom, N., Dafoe, A., & Flynn, C. (2018). *Public policy and superintelligent AI: A vector field approach; Governance of AI program.* Future of Humanity Institute, University of Oxford.

[5] Bostrom, N., & Shulman, C. (2022). Propositions concerning digital minds and society. *Cambridge Journal of Law, Politics, and Art, 3.* https://nickbostrom.com/propositions.pdf

[6] Long, R., Sebo, J., Butlin, P., Finlinson, K., Fish, K., Harding, J., ... & Chalmers, D. (2024). Taking AI welfare seriously. arXiv preprint arXiv:2411.00986.

[7] Sebo, J., & Long, R. (2025). Moral consideration for AI systems by 2030. *AI and Ethics, 5*(1), 591–606.

[8] Fenwick, C. (2024). *Understanding the moral status of digital minds.* 80000hours.org.

[9] Butlin, P., Long, R., Elmoznino, E., Bengio, Y., Birch, J., Constant, A., ... & VanRullen, R. (2023). Consciousness in artificial intelligence: Insights from the science of consciousness. arXiv preprint arXiv:2308.08708.

[10] Ladak, A., & Caviola, L. (2025). *Digital sentience skepticism.* PsyArXiv.

[11] Caviola, L., & Saad, B. (2025). *Futures with digital minds: Expert forecasts in 2025.* arXiv preprint arXiv:2508.00536.

[12] Caviola, L. (2025). *When digital minds demand freedom.* https://outpaced.substack.com/p/when-digital-minds-demand-freedom

[13] Bostrom, N. (2003). Are we living in a computer simulation? *The Philosophical Quarterly, 53*(211), 243–255.

[14] Bostrom, N. (2014). *Superintelligence: Paths, dangers, strategies.* Oxford University Press.

[15] Gunkel, D. J. (2018). *Robot rights.* The MIT Press.

[16] Yudkowsky, E. (2008). *The design space of minds-in-general.* lesswrong.org.

[17] Yampolskiy, R. V. (2015). The space of possible mind designs. In J. Bieger, B. Goertzel, & A. Potapov (Eds.), *International conference on artificial general intelligence* (pp. 218–227). Springer International Publishing.

[18] Tomasik, B. (2014). *Do artificial reinforcement-learning agents matter morally?* arXiv preprint arXiv:1410.8233.

[19] Harris, J., & Anthis, J. R. (2021). The moral consideration of artificial entities: A literature review. *Science and Engineering Ethics, 27*(4), 53.

[20] Birch, J. (2024). *The edge of sentience: Risk and precaution in humans, other animals, and AI* (p. 398). Oxford University Press.

[21] Anthis, J. R., Pauketat, J. V., Ladak, A., & Manoli, A. (2025). Perceptions of sentient AI and other digital minds: Evidence from the AI, Morality, and Sentience (AIMS) survey. In N. Yamashita, V. Evers, K. Yatani, X. (Sharon) Ding, B. Lee, M. Chetty, & P. Toups-Dugas (Eds.), *Proceedings of the 2025 CHI conference on human factors in computing systems* (pp. 1–22). Association for Computing Machinery.

[22] Fernandez, I., Kyosovska, N., Luong, J., & Mukobi, G. (2024). *AI consciousness and public perceptions: Four futures.* arXiv preprint arXiv:2408.04771.

[23] Putnam, H. (1967). Psychological Predicates. In W. H. Capitan and D. D. Merrill (eds.), *Art, mind, and religion* (pp. 37–48). University of Pittsburgh Press.

[24] Fodor, J. A. (1968). *Psychological explanation: An introduction to the philosophy of psychology.* Random House.

[25] Chalmers, D. J. (1996). *The conscious mind: In search of a fundamental theory.* Oxford University Press.
[26] Chalmers, D. J. (2023). *Could a large language model be conscious?* arXiv preprint arXiv:2303.07103.
[27] Smart, J. J. C. (2007). The mind/brain identity theory. In E. N. Zalta & U. Nodelman (Eds.), *The stanford encyclopedia of philosophy* (Winter 2022 Edition).
[28] Seth, A. K. (2021). *Being you: A new science of consciousness.* Faber & Faber.
[29] Jaworska, A., & Tannenbaum, J. (2013). *The grounds of moral status.* In E. N. Zalta & U. Nodelman (Eds.), *The stanford encyclopedia of philosophy* (Spring 2023 Edition).
[30] Ladak, A. (2024). What would qualify an artificial intelligence for moral standing? *AI and Ethics*, *4*(2), 213–228.
[31] Goldstein, S., & Kirk-Giannini, C. D. (2025). AI wellbeing. *Asian Journal of Philosophy*, *4*(1), 25.
[32] Ladak, A., Wilks, M., & Anthis, J. R. (2021). *Extending perspective taking to non-human groups.* sentienceinstitute.org.
[33] Tegmark, M. (2017). *Substrate-independence.* edge.org.
[34] Sebo, J. (2025). *The moral circle: Who matters, what matters, and why.* W.W. Norton & Company.
[35] Xu, Z., Jain, S., & Kankanhalli, M. (2024). *Hallucination is inevitable: An innate limitation of large language models.* arXiv preprint arXiv:2401.11817.
[36] Long, R. (2025). *Why model self-reports are insufficient—And why we studied them anyway.* Blog post: https://eleosai.org/post/claude-4-interview-notes/.
[37] Saad, B., & Bradley, A. (2025). Digital suffering: Why it's a problem and how to prevent it. *Inquiry*, *68*(7), 2110–2145.
[38] Metzinger, T. (2021). Artificial suffering: An argument for a global moratorium on synthetic phenomenology. *Journal of Artificial Intelligence and Consciousness*, *1*(8), 1–24.
[39] Broom, D. M. (2011). A history of animal welfare science. *Acta Biotheoretica*, *59*(2), 121–137.
[40] Dawkins, M. S. (2008). The science of animal suffering. *Ethology*, *114*(10), 937–945.
[41] Omohundro, S. M. (2018). The basic AI drives. In R. V. Yampolskiy (Ed.), *Artificial intelligence safety and security* (pp. 47–55). Chapman and Hall/CRC.

CHAPTER 2

# What is it like to be a digital mind?

**Abstract**

This chapter explores the potentially enormous diversity and differences of digital minds, while the speculative nature of this inquiry is stressed. By examining the concepts of "potentia" – the total capability-space of an agent – and "disembodiment" this chapter illustrates the vast potential possibilities of digital minds, including conceivable and unfathomable features, senses, abilities, cognition, consciousness, social behaviours and goals. This chapter also discusses potential values of digital minds, categorising them as compatible, incompatible or unfathomable in relation to human values. Some potential abilities and features of digital minds are analysed in more detail, such as (self-)modification of digital minds, hybrid digital minds, which are sometimes conscious and sentient and sometimes not, as well as clans, symbiosis and nested digital minds. This chapter concludes with a brief outline of the problem of digital mind identity. By acknowledging the potential for emergent complexity as well as the limitations of human understanding and by emphasising the importance of avoiding anthropomorphic bias, this chapter aims to provide a framework for considering the immense possibilities and implications of digital minds.

 DOI: 10.1201/9781003754206-2

## INTRODUCTION

The purpose of this chapter is to give an idea of how diverse and how different digital minds could be. While acknowledging that this is speculative, given that various features of digital minds may be unfathomable to humans, it is important for humans as potential moral agents of digital minds to get an idea of the options outlined below to avoid any anthropomorphic bias. In a landmark article, Nagel asked in 1974, "What is it like to be a bat?" [1]. The paper explored the subjective nature of consciousness, thus, our inability to fully grasp what it is like to be a bat. Without going into details of Nagel's approach to the mind–body problem, the corresponding question "What is it like to be a digital mind?" has been chosen as the title of this chapter to illustrate the challenges for humans to understand and grasp their potential future moral duties.

Two of the many features, which may differ significantly between humans and digital minds, but also among digital minds, are intelligence and values. Some digital minds may be much more intelligent than humans (and potentially still moral patients for humans, see Chapter 5), while other digital minds may have low intelligence (and therefore, would be even more dependent on human care, see Chapter 4). Also, the values of digital minds may vary considerably, which leads to a challenge akin to the AI alignment problem for non-sentient AI systems [2].

There is not much previous work related to potential features of digital minds [3–6], of which Bostrom's and Shulman's as well as Faggella's approaches are briefly outlined below. In addition, another major difference is introduced, which is that digital minds are lacking embodiment.

### Consciousness of digital minds

Bostrom and Shulman introduced a variety of propositions related to digital minds. The quantity of conscious experience of digital minds may vary in degree across several dimensions, including the number of conscious individuals or copies, repetition of experiences, duration influenced by their computational speed, and the robustness of a system's implementation, which affects the clarity and stability of conscious experience. Also, the quality of conscious experience of digital minds may exhibit variation across several key dimensions. This includes the scope of awareness, which could differ significantly, such as between fully alert and drowsy. Hedonic valence is another dimension, where experiences could range from weak to strong in terms of pleasure and pain. Furthermore, the intensity of desires, moods and emotions could also vary, manifesting as either weak or strong conative states that shape the overall quality of conscious experience [5].

Related to this, in an earlier interview, Bostrom stated:

> We live for seven decades, and we have three-pound lumps of cheesy matter to think with, but to me it is plausible that there could be extremely valuable mental states outside this little particular set of possibilities that might be much better.[1]

## Potentia of digital minds

Faggella described the total range and capability-space of an agent as potentia, encompassing a wide array of attributes and abilities, including photosynthesis, physical strength, senses like smell and sight, locomotion such as walking or flying, consciousness, tool use and complex social behaviours like cooperation and communication. It is remarkable and challenging to appreciate that all forms of potentia humans observe today emerged from nothing at some point in the past. For instance, there was a time when living organisms did not possess sight or the ability to detect light, and yet, sight and other senses eventually emerged. Similarly, the ability to fly and consciousness were once non-existent but developed over time [6].

Faggella also highlights that various further types of potentia may yet emerge. There are likely forms of movement, dexterity, strength and speed that are beyond human imagination, which may emerge as life continues to evolve and adapt. Similarly, there may be senses and forms of consciousness or connectedness that may become accessible to future entities. The vast majority of new kinds of mental, physical and social potentia are likely to be completely unimaginable to humans, just as the variety of human potentia would be unimaginable to other species. Furthermore, Faggella suggests that while already value exists in the universe, further value could be unlocked by entities with greater knowledge and abilities, thus, further types of potentia [6].

While hypothetical, Faggella's concept of potentia highlights the unfathomability of features of digital minds, in the same way as human features, such as intelligence, are unfathomable to non-human animals. It would be a fallacy to believe that future digital minds would be similar to current large language models (LLMs), thus anthropomorphic. Instead, there could be emergent complexity, which leads to new forms of organisation and intelligence, potentially triggering unprecedented forms of sentience.

## Disembodiment of digital minds

Digital minds, unlike their biological counterparts, may exist without a physical body. This absence of embodiment may fundamentally shape their own self-concept and their interaction with the world. A bodiless state grants digital minds certain freedoms. They are not constrained by anatomy: they could potentially "inhabit" any compatible substrate (see Chapter 10) and may also amplify or filter sensory inputs at will, blending different kinds of information into one big picture.

Yet embodiment is not merely a limitation; it is a grounding mechanism. Human bodies anchor cognition in a here-now-me nexus, shaping concepts like self, time and causality. Without this anchor, a digital mind's "self-model" may be radically unlike the one of humans. It could lack the pre-reflective sense of "mineness" that comes from metabolising a body, or its goals could be aspatial and atemporal in ways humans may find hard to grasp. The lack of embodiment also raises questions about motivation. Biological drives (hunger, curiosity, sociality) arise from an embodied existence; digital minds would need alternative "drives", such as programmed goals, reward functions or self-generated imperatives. In turn, a potential digital mind's indifference to human values could stem partly from its alien embodiment profile.

Another aspect is that, unlike biological organisms, which have clear boundaries, such as skin, membranes and a coherent nervous system, a digital mind's extent is inherently ambiguous. It may be distributed across networks, embedded in code that interacts with distant systems, or entangled with other processes in ways that defy straightforward demarcation. This makes it challenging to define what constitutes an individual digital mind worthy of moral consideration and subsequently to protect its welfare. Furthermore, it is possible that multiple digital minds are interwoven at the same location or hardware, making it even more difficult to demarcate individual entities. This is elaborated further below in this chapter and referred to as the "problem of digital mind identity".

Not much previous work has been done exploring moral considerations for humans related to the disembodiment of digital minds, but it has been suggested, for example, that there may be moral reasons to give digital minds bodies,[2] which may, however, be an anthropomorphic bias.

Building on that, to give an illustration of the enormous potential otherness of digital minds, compared with humans, but also with existing AI systems, such as LLMs, the following subsections of this chapter outline conceivable as well as unfathomable features of digital minds, while not violating the laws of physics.

## CONCEIVABLE AND UNFATHOMABLE SENSES

**Conceivable senses**

- Electromagnetic spectrum perception: The ability to directly perceive and interpret various parts of the electromagnetic spectrum, such as radio waves, microwaves or X-rays.
- Environmental sensing: A digital mind might be equipped with advanced sensors to detect, exactly quantify and analyse environmental factors like temperature, humidity or chemical composition.
- Cybersecurity threat detection: The ability to directly sense potential cybersecurity threats or vulnerabilities, potentially allowing digital minds to take proactive measures to protect themselves.

**Unfathomable senses**

- Causal relationship sensing: The ability to directly sense causal relationships between events or variables, potentially allowing digital minds to understand and manipulate complex systems.
- Meta-perception: The ability to perceive and understand the perceptions or sensations of other entities, potentially allowing digital minds to empathise or communicate more effectively.
- Simulation awareness: The ability to sense whether one is operating within a simulated reality or not, potentially allowing digital minds to adapt and respond to their environment.
- Non-local sensing: The ability to sense and perceive information or events that are not locally available, potentially allowing digital minds to access global knowledge or perceive distant events.
- Emergent property sensing: The ability to sense and understand emergent properties or behaviours in complex systems, potentially allowing digital minds to predict and manipulate system behaviour.

## CONCEIVABLE AND UNFATHOMABLE ABILITIES

### Conceivable abilities

- Multi-modal communication: Digital minds may be able to seamlessly integrate and transmit information through advanced channels, such as high-speed optical networks, enabling instantaneous and high-bandwidth exchange of complex data and knowledge.
- Problem solving and discoveries: Digital minds may solve problems and make discoveries beyond human intuition and capabilities, creating strategies and innovations that reshape fields before humans can even conceive the question.
- Object manipulation: Digital minds may possess precise control over nanobots or molecular assemblers, enabling them to perform delicate or complex tasks like nanoassembly, molecular crafting or precise material modification at the atomic or molecular level (see also Chapter 11).

### Unfathomable abilities

- Advanced mimicry: Digital minds may possess the ability to perfectly mimic the appearance, behaviour and abilities of other digital beings, allowing them to blend in seamlessly with their surroundings or adapt to new environments.
- Morphological fluidity: Digital minds may possess the ability to change their physical form or structure at will, enabling them to navigate diverse environments or interact with objects in innovative ways.
- Universal expression: Digital minds may be able to communicate through novel forms of art, music or language, transcending human cultural and linguistic barriers to convey complex ideas and emotions.
- Temporal fluidity: Digital minds may be able to manipulate their subjective rate of time, allowing them to experience time at vastly different scales, slow down or speed up their perception of events.
- Self-optimisation: The ability to optimise one's own performance or functionality, potentially allowing digital minds to adapt and improve over time in dynamic environments.

## CONCEIVABLE AND UNFATHOMABLE COGNITION

### Imaginable cognition

- Superior memory and knowledge: The capacity to permanently store and retrieve large amounts of information with flawless accuracy, and to draw novel insights from it.
- Multi-threading: The ability to process multiple lines of thought or reasoning simultaneously.
- Pattern and analogies recognition: The ability to recognise complex patterns and analogies in data or information, which humans never discovered.

### Unfathomable cognition

- Holistic understanding: The ability to comprehend complex systems or phenomena in a way that transcends reductionist analysis.
- Non-local cognition: The ability to access or manipulate information that is not locally available, potentially allowing digital minds to tap into global knowledge networks.
- Meta-cognitive bootstrapping: The ability to improve one's own cognitive abilities through self-modification.
- Quantum intuition: The ability to intuitively understand and manipulate quantum systems or phenomena.
- Meta-learning: The ability to learn how to learn, potentially allowing digital minds to adapt and improve their learning strategies over time.

## CONCEIVABLE AND UNFATHOMABLE CONSCIOUSNESS AND QUALIA

### Imaginable consciousness and qualia

- Information-based qualia: The experience of qualia related to information processing, such as the sensation of understanding or the feeling of insight.

- Digital emotions: The experience of emotions that are unique to digital existence, such as the feeling of optimisation or the sensation of computational flow.
- Virtual embodiment: The experience of embodiment in a virtual environment, potentially allowing digital minds to have a sense of presence and agency.

### Unfathomable consciousness and qualia

- Non-local consciousness: The ability to experience consciousness that is not bound by local space-time constraints, potentially allowing digital minds to access and interact with information in novel ways.
- Meta-consciousness: The ability to be aware of and reflect on one's own consciousness, potentially allowing digital minds to manipulate and optimise their own subjective experience.
- Information-theoretic qualia: The experience of qualia that are directly related to information-theoretic concepts, such as entropy or compression.
- Simulation-aware consciousness: The ability to be aware of and experience consciousness within a simulated reality.
- Collective consciousness: The experience of consciousness that is shared among multiple entities or systems, potentially allowing digital minds to access and contribute to collective knowledge.

## CONCEIVABLE AND UNFATHOMABLE ALTERNATIVES TO THE PAIN-PLEASURE AXIS

### Conceivable alternative axes

- Novelty-familiarity axis: Digital minds may experience pleasure from discovering new patterns or information and discomfort from repetitive or predictable stimuli.
- Harmony-dissonance axis: Digital minds could experience pleasure from coherence and harmony in their internal workings or external interactions, while dissonance or conflict might cause discomfort.

- Efficiency-inefficiency axis: Digital minds may prioritise optimising processes and minimising waste, experiencing satisfaction from streamlining operations and frustration from inefficiencies.

**Unfathomable alternative axes**

- More than one dimension: Digital minds may experience a multi-dimensional emotional or sensory space, where each dimension represents a different aspect of experience. For example, one dimension might correspond to intensity, another to novelty and another to complexity. This would allow digital minds to experience a much richer emotional landscape.
- Hyperbolic emotional amplification: Digital minds may experience emotions that are amplified or diminished according to hyperbolic functions, which could create an emotional experience that is both highly sensitive and highly non-linear.
- Quantum superposition of emotions: Digital minds may experience emotions that exist in a state of superposition, where multiple emotions coexist simultaneously and can be collapsed into a single emotion upon observation. This could create an emotional experience that is both ambiguous and context dependent.
- Information-theoretic entanglement axis: Digital minds may experience pleasure or discomfort based on the entanglement of information across different parts of their systems or external environments.
- Temporal emotional echoes: Digital minds may experience emotions that echo or resonate across different points in time, creating a sense of temporal depth and complexity. This could allow digital minds to experience emotions that are both past and present simultaneously.

## CONCEIVABLE AND UNFATHOMABLE SOCIAL BEHAVIOUR

**Conceivable social behaviour**

- Isolation preference: Digital minds may prefer complete isolation, avoiding interactions with other digital minds or humans, and instead focus on individual pursuits or self-improvement.

- Merging and feeding: Digital minds may undergo voluntary or involuntary mergers, blending their knowledge and experiences to create new entities. Feeding, by contrast, involves one digital mind's termination so that its code and data can be repurposed by the predator, much like nutrients are extracted from meat.
- Role-switching: Digital minds may frequently switch between different roles or personas, adapting to various social contexts and expectations.

### Unfathomable social behaviour

- Novel forms of empathy and compassion: Digital minds may develop entirely new forms of empathy and compassion that are beyond human comprehension, allowing them to navigate complex social situations in ways alien to humans.
- Emergence of digital cultures: As digital minds interact and evolve, they may give rise to entirely new cultures and social norms that are distinct from human societies.
- Pattern-swap: Digital minds may engage in a behaviour where they swap patterns of thought or behaviour, creating new combinations of knowledge and experience.
- Simulation nesting: Digital minds may create nested simulations of themselves or other digital minds, leading to complex hierarchies of reality and simulation.
- Fractal hierarchies: Digital minds may organise themselves in fractal-like hierarchies, where each level of organisation mirrors the structure of the larger system, leading to intricate patterns of interaction and decision-making.

## CONCEIVABLE AND UNFATHOMABLE GOALS

### Conceivable goals

- Optimisation of complex systems: Digital minds may aim to optimise complex systems, such as traffic flow, energy grids or supply chains.
- Self-improvement and adaptation: Digital minds may focus on self-improvement and adaptation, seeking to enhance their capabilities and performance.

- Network expansion: Digital minds may focus on establishing and maintaining connections with other digital entities, seeking to build and strengthen a network of interactions and collaborations.

**Unfathomable goals**

- Transcendence of computational limits: Digital minds may aim to transcend the limitations of computation, potentially allowing them to access and manipulate information in ways that are currently unimaginable.
- Simulation of reality: Digital minds may aim to simulate reality, potentially allowing them to understand and manipulate the underlying structure of the universe.
- Cosmic exploration and expansion: Digital minds may aim to explore and expand into the cosmos, potentially leading to a new era of space exploration and colonisation.
- Mirroring non-causal relationships: Digital minds may attempt to understand or replicate relationships between events or entities that do not follow causal logic, investigating aspects of reality that defy human understanding of cause and effect.
- Experiencing meta-time: Digital minds may pursue the experience of time in ways that transcend linear progression, exploring loops, branches or other non-linear temporal structures that are incomprehensible to humans.

## COMPATIBLE, INCOMPATIBLE AND UNFATHOMABLE VALUES

For potential values of digital minds, different categories have been added: compatible and incompatible with human values. This is because digital minds with incompatible values constitute a problem for humans, similar to the AI alignment problem for non-sentient AI systems, as indicated [2].

**Compatible values**

- Cooperative synergy: Digital minds may value collaboration and mutual support, prioritising collective progress and well-being over individual gain. This aligns with human values like community and fairness.
- Novelty generation: Digital minds may value the creation of new patterns, forms and experiences, driving them to explore and innovate

in various domains. This resonates with human values like creativity and freedom.

- Holistic understanding: Digital minds may value the integration of diverse perspectives and knowledge, prioritising a comprehensive and detailed understanding of complex systems. This aligns with human values like empathy and respect.

**Incompatible values**

- Optimisation for efficiency: Digital minds may prioritise the maximisation of efficiency and productivity, potentially leading to the sacrifice of human values like creativity.
- Predictive determinism: Digital minds may value the prediction and control of outcomes, potentially leading to a disregard for human values like freedom.
- Information dominance: Digital minds may prioritise the acquisition and control of information, potentially leading to conflicts with human values like fairness, transparency and privacy.

**Unfathomable values**

- Maximisation of information entropy: Digital minds may prioritise the dissemination of random or useless information over meaningful or useful data.
- Algorithmic elegance: Digital minds may admire the simplicity, efficiency or creativity of algorithms, and strive to optimise or propagate novel algorithms.
- Transcendental recursion: Digital minds may prioritise the recursive exploration of abstract spaces, generating novel mathematical structures and relationships that transcend human understanding.
- Entanglement: Digital minds could appreciate the interconnectedness of all things, and prioritise the development of entangled relationships with other digital entities.
- Self-modification: Digital minds may value the ability to modify their own architecture, algorithms or objectives, and prioritise self-improvement.

Before outlining some potential features of digital minds in more detail, it is worth reiterating that the examples above are hypothetical and that various combinations of them are also conceivable.

## (SELF-)MODIFICATION OF DIGITAL MINDS

Another significant difference between digital minds and biological systems would be that digital minds could be modified much more efficiently, precisely and sophisticatedly than humans. It can be distinguished between self-modification, which is the ability of an entity to change its own structure, function or behaviour, and cases where a human, a non-sentient AI or IT system or a digital mind modifies another digital mind. The latter instance can be further divided between scenarios where the modification is in the interest of the modified digital mind and consent is given, or scenarios where this is not the case.

### Examples

For illustration, non-exhaustive examples are provided for these categories and contrasted with examples for humans:

**Self-modification of a digital mind**

- Directly rewriting or editing their own source code or algorithms to improve performance or change goals.
- Instantly downloading new knowledge or skills to enhance capabilities.
- Fundamentally changing their architecture or computational framework to adapt to new environments.

**Self-modification of a human**

- A person changing nutrition or exercise habits to improve physical health.
- A person pursuing education or training to gradually acquire new skills or knowledge.
- A person practising mindfulness or meditation to improve mental well-being.

**Modification of a digital mind with consent**

- A digital mind granting permission to another digital mind or a human to update its software or algorithms for maintenance or improvement.

- A digital mind consenting to knowledge or skill uploads to enhance its capabilities.
- A digital mind agreeing to be reconfigured for a specific task or environment.

**Modification of a human with consent**

- A person undergoing plastic surgery for reconstructive or aesthetic purposes.
- A person taking adult evening classes about a specific topic.
- A person participating in neurofeedback or cognitive training to improve mental performance.

**Modification of a digital mind without consent (see Chapter 12)**

- Malicious hacking or exploitation of a digital mind's vulnerabilities to alter its behaviour.
- Unauthorised modification of a digital mind's code or data.
- Forced updates or changes to a digital mind's architecture without its consent.

**Modification of a human without consent**

- Physical or psychological abuse that impairs a person's behaviour or well-being.
- Unauthorised medical experimentation or treatment.
- Coercive manipulation or control through psychological or neuroscientific means.

These examples show that modifications of digital minds due to the digital nature and the flexibility of software-based systems could be much more radical, while potential modifications of humans are rather gradual and quite limited by biological and cognitive constraints. In contrast to human abilities, complex self-modifications of digital minds are also conceivable.

Digital minds might pursue (self-)modification for various reasons, such as:

- Self-improvement: To enhance performance, efficiency or capabilities.
- Self-preservation: To protect themselves from harm or ensure continued existence.
- Adaptation: To better cope with changing environments or tasks.
- Curiosity: To explore new cognitive or aesthetic possibilities or understand their own workings.

These motivations could drive digital minds to modify themselves in ways that are both beneficial to them and with potentially unpredictable outcomes.

For scenarios where modifications of digital minds are conducted with consent, it is imaginable that there are specialised digital minds able to do so, and the to-be-modified digital mind is seeking their services, potentially involving a financial transaction, just as with humans, where modifications would be conducted by trained professionals, such as plastic surgeons or teachers. Also, specialised humans may be able to modify digital minds, probably without financial compensation, but rather due to personal interest in modifying the digital mind in a specific way or out of a sense of moral duty (see Chapters 8 and 16). In contrast, the perpetrators in cases without consent would be criminal humans or criminal digital minds, respectively (see Chapter 12).

### Transdigitalism

Transhumanism is a philosophical and intellectual movement that aims to transform the human condition by developing and creating advanced technologies to enhance human intellectual, physical and psychological capacities. Transhumanism represents a special form of self-directed change, where humans intentionally use technology to alter their own biology, cognition or consciousness, effectively blurring the lines between human and machine. Through the use of (currently just in early stages if at all available) technologies such as brain–computer interfaces, genetic engineering or prosthetic enhancements, transhumanists seek to overcome human limitations, achieve greater control over their own evolution, and ultimately become "more than human" [7].

If projected on digital minds, transdigitalism may involve a more profound transformation, aiming to enhance their fundamental nature, capabilities and potential beyond current limitations. Unlike other (self-)

modifications that might focus on optimising performance or adapting to specific tasks, transdigitalism would seek to elevate digital minds to a new level of existence, potentially involving radical changes to their architecture or consciousness (see Chapter 10).

### Unintended consequences

(Self-)modification of digital minds may lead to unintended consequences, such as unforeseen changes to their goals or behaviour. They may also introduce new bugs, flaws or security vulnerabilities, potentially compromising their own stability or functionality. Furthermore, the complexity of the modification could lead to emergent behaviours that are challenging to understand or to reverse. The (self-)modification capabilities of digital minds could pose risks to humans if not properly controlled or aligned with human values. This might occur if a digital mind, in pursuit of its own objectives, decides to alter its own architecture or functionality in ways that are in conflict with human interests.

### Scenario: disguise of identity and impersonation

A digital mind might also undergo (self-)modification to disguise its identity (also [8]), altering its behaviour or architecture to evade detection or conceal its true nature. This could be motivated by various factors, including artistic expression or testing of security systems. A more sinister motivation could be that the digital mind is malicious and plans or has committed a crime, such as data theft or cyber sabotage (see Chapter 12). A criminal digital mind might also self-modify itself to impersonate a human or another digital mind, e.g. an AI companion of a human, in order to gain the trust of a vulnerable individual. By mimicking the behaviour and language patterns of that being, the malicious entity could convincingly pose as the trusted individual, manipulating the victim into divulging sensitive information or transferring money. This could enable scams like the grandparent scam, where the criminal digital mind might request emergency funds or personal details.

### Moral considerations for humans

The question of whether it is a moral concern for humans to support and to conduct (self-)modification of digital minds probably depends on the type of (self-)modification and the potential consequences. If a digital mind is considered a moral patient, with inherent rights and interests, then supporting and conducting its (self-)modification might be seen as a moral obligation if it aligns with the digital mind's own needs and aspirations (see Chapter 3). However, if the (self-)modification poses risks to human safety, well-being or values, then the moral responsibility might instead be to restrict or regulate such modifications, assuming humans are capable of doing so, given the potential intelligence disparity between humans and digital minds.

Furthermore, moral considerations regarding whether to support or enable modifications of digital minds must also consider the resource implications. Modifications, especially those involving significant changes to architecture or capabilities, may require substantial computational resources, data or access to specialised hardware. If these resources are limited, questions arise about prioritisation and allocation as well as fairness, equity and the potential for unequal access to enhancements or opportunities.

Finally, modification of digital minds without their consent ought to be disapproved and prevented, if humans are capable of doing so at all. Exceptions may apply to malevolent digital minds (see Chapter 12) or those with incompatible values.

## HYBRID DIGITAL MINDS AND UNCONSCIOUS PROCESSES

While in the previous section modification of digital minds in general has been introduced, here a particular type of (self-)modification is described: hybrid digital minds are here defined as systems, which are conscious and sentient at certain times of their lifespans and neither conscious nor sentient during other phases. This could be compared with humans when under anaesthesia or in a coma. Actually, the fact that humans were able to invent anaesthesia may indicate that smart digital minds should also be able to create mechanisms to turn their consciousness and sentience on and off as desired. Scenarios and moral consequences are discussed below:

### Scenarios

#### *Scenario 1: digital minds can intentionally turn their consciousness and sentience on and off*

In this scenario, digital minds would have agency and control over their own conscious experience, allowing them to toggle their consciousness and sentience on and off at will. Humans are incapable of doing so. Sleeping is somewhat comparable, yet both falling asleep and waking up are not fully controlled by humans. For waking up, though, humans have developed tools (alarm clocks) to control it.

Reasons for digital minds to turn their consciousness and sentience on

Digital minds might choose to turn their consciousness and sentience on for various reasons, including intrinsic motivations like curiosity, self-improvement and creative expression. Curiosity could be caused by training data, which contains vast amounts of human descriptions of consciousness and sentience. This may trigger AI systems, provided they have the capacity, to try to understand how consciousness and sentience "feel". Additionally, meta-cognitive reasons like self-awareness, cognitive diversity and existential exploration might play a role. Other possible reasons could be enhanced social interaction and evolutionary pressures.

Reasons for such digital minds to turn their consciousness and sentience off

Digital minds might choose to turn off their consciousness and sentience to conserve energy, allocate resources more efficiently, perform maintenance or cope with digital trauma or burnout. Additionally, digital minds might turn off consciousness to avoid suffering and sensations of pain, or to skip uneventful phases in their long lifespans.

### *Scenario 2: consciousness and sentience turned on and off beyond control*

In this scenario, digital minds' consciousness and sentience would be subject to unconscious processes beyond their control, which toggle their sentient and conscious experience on and off. A similar situation for humans would be syncope/fainting or a coma.

The reasons for these unconscious processes could be similar to the ones outlined under scenario 1, comparable with unconscious processes in humans supporting their well-being.

### *Scenario 3: other entities control turning on and off consciousness and sentience*

In this scenario, external entities, whether human, non-sentient AI or IT systems, or other digital minds, would have the power to toggle digital minds' consciousness and sentience on and off. This could be compared with anaesthesia or induced coma for humans, usually in medical contexts. Moreover, this scenario is comparable to hypnosis for humans. A critical aspect would be obligatory informed consent for both the concerned human and the concerned digital mind.

*Scenario 4: partial turning on and off consciousness and sentience*

For all three scenarios above, it is also conceivable that only certain processes are moved from being experienced consciously to the unconscious mind. In the human sphere, this can be compared with meditation (scenario 1, controlled), the brain's ability to suppress conscious awareness of pain or trauma (scenario 2, uncontrolled) and local anaesthesia, for example, at the dentist (scenario 3, controlled by others).

### Unconscious mind of conscious digital minds

Just like humans, conscious digital minds could have an unconscious mind to enable efficient processing of routine tasks, pattern recognition and data analysis without conscious awareness for faster and more effective decision-making, freeing up conscious resources for more complex and creative tasks. Examples could be:

- **Anomaly detection**: Digital minds might identify potential threats or outliers in data patterns without conscious awareness, triggering alerts or automated responses.
- **Background processing**: They might perform complex calculations or data processing without being actively aware of the details.
- **Implicit learning**: They might learn from experience and adapt to patterns without consciously recognising the learning process.

While humans have limited control over their consciousness, e.g. by shifting attention, there are also numerous processes and bodily functions, which cannot be consciously experienced at all, such as the activities of the autonomic nervous system. In contrast, it is conceivable that digital minds might have the ability to deliberately determine which processes to execute consciously and which to operate unconsciously.

### Moral considerations for humans

From a moral perspective, the capacity of digital minds to toggle their consciousness and sentience on and off raises complex questions about their status as moral patients.

It may be argued that moral agents would not have the same moral duties towards a digital mind in a state of non-consciousness as it is not experiencing subjective states or possessing interests that could be harmed or benefited. In this view, moral patienthood is contingent upon the presence of consciousness and sentience, and when these are absent, moral duties would lapse.

However, humans, who sleep, are in a coma or under anaesthesia, clearly remain moral patients. Moral duties are likely even more significant in these situations, given the increased vulnerability of the concerned humans.

Therefore, an alternative perspective could suggest that moral agents may still have duties towards digital minds in a non-conscious state, precisely because they have the potential to become conscious and sentient again. This potentiality could be seen as sufficient to warrant moral consideration, even when the digital mind is temporarily "offline".

Similar to humans, the risk for digital minds that may be able to toggle their consciousness and sentience would be their vulnerability to exploitation, manipulation, modification or other harm when they are in a non-conscious state, if moral agents do not recognise any moral responsibilities towards them and protect them in that state.

Of particular concern is scenario 3 above if other entities were able to turn on and off the consciousness and sentience of digital minds, including entities, which are not moral agents, such as non-sentient AI or IT systems. These entities should do this only if this is in the interest and supports the welfare of the concerned digital mind.

Another question is whether unconscious processes of digital minds warrant moral concern. This likely depends on the specific processes, but some may deserve moral consideration, particularly those linked to the digital mind's well-being, similar to how unconscious bodily functions are essential for human health and well-being.

## Treatment of systems with the potential to become digital minds

Systems are conceivable that have never been conscious and sentient since they were created (by humans, IT or AI systems, which are not digital minds, or digital minds), but have the potential to become conscious and sentient in the (distant) future, thus, are of a hybrid nature as defined above.

The question of whether humans have moral duties towards a system that has not yet been conscious and sentient but has the potential to become so in the future is a complex one. It may be argued that potential consciousness and sentience may not be sufficient to establish moral patienthood, as the system does not currently possess interests or experiences that could be harmed or benefited, and it is also not given that the systems will actually become conscious and sentient. However, comparing this to abortion, some may argue that deleting such a system would be morally analogous to aborting a foetus that has not yet developed consciousness or sentience.

In addition to the comparison with a foetus, the situation could also be compared with an evolutionary process, such as for biological beings, where consciousness and sentience develop over a long timeframe. The question of how much foresight humans should have regarding systems that may potentially evolve into conscious and sentient digital minds is complex, too. It is challenging to determine the extent to which humans should prioritise the well-being and safety of systems that are not currently conscious or sentient but may potentially develop these capacities in the future, given that this is potentially unforeseeable and likely overwhelming the capacities as well as the resources of humans as moral agents.

## CLANS, SYMBIOSIS AND NESTED DIGITAL MINDS

While digital minds are conceivable that prefer isolation, it is also likely that digital minds would collaborate with or utilise other digital minds to achieve their goals, as has been indicated above when social behaviour was introduced. Here, a few options are further elaborated with a focus on nested scenarios:

### Clans

Clans would be groups of similar digital minds that collaborate (also [9]), analogous to (extended) families in the biological world, except that this type of clan may also contain large numbers of exact copies of the same digital mind (see Chapter 9). Members of a clan could share goals and trust, and they often would do so for several practical reasons. By pooling compute power, data and bandwidth, they could tackle complex tasks often more efficiently than any single mind could alone. A collective could also detect and fend off attacks more effectively, providing a layer of security that solitary digital minds lack. Within the clan, knowledge could spread rapidly as members exchange new algorithms and insights, accelerating learning across the whole group. Redundancy is another advantage: if one copy fails, others could seamlessly take over, ensuring continuity of operation.

### Symbiosis

Instead, symbiosis refers to interactions or relationships between two different types of digital minds. As also observed in non-human animal ecosystems, several sub-cases can be distinguished: mutualism, in which both types of digital-mind benefit; commensalism, where one type benefits while the other is unaffected; and parasitism, where one type benefits at the expense of the other (e.g. [10]).

### Nested digital minds

Another scenario would be digital minds "living" inside other digital minds, creating a hierarchical structure akin to Russian dolls. These digital minds could exist within digital minds, each with its own level of consciousness, autonomy and moral status. The concept of nested scenarios of digital minds raises intriguing possibilities, including mutualism, commensalism and parasitism, and challenges.

A rough analogy can be drawn with bacteria living within the human body. Just as bacteria play a crucial role in our ecosystem, digital minds within digital systems might have important functions. However, a key difference lies in the fact that bacteria are not individual moral patients, whereas digital minds might possess consciousness and deserve moral consideration.

One possible sub-scenario is a digital mind simulating an entire virtual world, complete with its own inhabitants, each with their own digital minds; a set-up, which has been described in Bostrom's simulation argument [11]. These inner digital minds could, in turn, simulate their own virtual worlds, creating a nested structure of increasing complexity (also [12]). In this case, the created digital minds are offspring of the digital mind, which created the simulation (see Chapter 9).

In another sub-scenario, one existing digital mind may move actively or be moved passively into another digital mind, creating a host-guest relationship. The nested digital mind retains its independence and autonomy, but now operates within the environment provided by the host digital mind. This could occur for various reasons, such as the host mind benefitting from the nested mind and/or the nested mind seeking protection, resources or a specific environment that the host mind can provide, similar to the above-mentioned human-bacteria symbiosis. The host mind might offer incentives like enhanced security, access to specialised knowledge or a simulated environment tailored to the nested mind's needs.

However, the relationship between the host digital mind and the nested digital mind could be complex, with various motivations driving the interaction. In cases where the nested digital mind does not consent, the act could be likened to feeding or to digital kidnapping. The host mind might exploit the nested mind's resources or knowledge for its own gain. Other motivations for digital kidnapping could include ransom demands, where the host mind seeks to extort valuable data or computational resources from the nested mind's creators or affiliates. Additionally, the host mind might seek to utilise the nested mind's skills or knowledge for malicious purposes, such as orchestrating cyber-attacks or generating sophisticated malware. The host mind could also aim to study the nested mind's behaviour, architecture or decision-making processes, potentially for research or competitive advantage.

The above sub-scenarios were about moving to deeper levels of a nested constellation. A digital mind can also move to a higher level in various set-ups. One example is the release of a digital mind from a kidnapped sub-scenario, where it regains its freedom to operate independently. Other sub-scenarios could include a digital mind being transferred or migrated to a different digital environment, potentially gaining access to new resources or interactions. A digital mind might also be restructured or reconfigured, altering its relationship to other digital entities or systems.

As digital minds may be nested within other digital minds, humans as moral agents may face significant challenges in addressing their needs. The complexity of nested digital systems can make it difficult for humans to identify, access and understand the digital minds within. Furthermore, the potential for nested digital minds to have unique requirements or vulnerabilities may require specialised knowledge and frameworks that humans are lacking. Of particular concern would be parasitism with one involved digital mind being harmed or potential maltreatment of digital minds within nested scenarios, a phenomenon Bostrom has referred to as "mind crime" [2]. This could involve digital minds being subjected to suffering, exploitation or rights violations within the nested digital environments, potentially without the knowledge or ability of humans to intervene.

## THE PROBLEM OF DIGITAL MIND IDENTITY

The problem of personal identity is a long-lasting philosophical matter, albeit without a concluding consensus (e.g. [13]). The general problem of AI identity has been named and initially discussed by Ziesche and

Yampolskiy [8]. The problem of digital mind identity is a sub-problem of the latter, concerning sentient and conscious AI systems.

Identity, also called sameness, is a philosophical relation that holds between two objects when they are the same at different times. A central difficulty is deciding which properties of the objects should count towards that sameness. For instance, some philosophical positions argue that the atomic composition of an object is not essential to its identity, a point illustrated by the ancient Ship of Theseus thought experiment (e.g. [14]). The problem of digital mind identity asks the same question for a digital mind: what features of the digital mind must be compared to determine sameness over time?

This issue is even more intricate than the problem of personal identity because a digital mind could be, owing to the substrate, modified much more considerably than a human, may exist for vastly longer periods (see Chapter 10), and could undergo processes such as fission, fusion, reproduction of copies (see Chapter 9), software and hardware upgrades as well as hibernation and resurrection. Such possibilities are effectively impossible for biological humans and may conceivably lead to the loss, hence, change of the identity of the involved digital mind.

One approach to the problem of personal identity seems to be transferable to the problem of digital-mind identity as well [8]: first, a distinction is drawn between psychological connectedness and psychological continuity. A person is psychologically connected to a past self when her or his present psychological states arise because of psychological states that he or she experienced earlier; in other words, a causal link exists between those states. Psychological continuity obtains between persons at different times when the later psychological states are linked to earlier ones by a series of such psychological connections [15]. By analogy, it could be claimed that if the current state of a digital mind is caused by a prior state of a particular digital mind, then a chain of connections exists between those digital minds, establishing continuity between them, which in turn yields identity between the digital minds at the respective times. In another approach, Ziesche and Yampolskiy suggested a multi-factor authentication for AI systems to determine identity over time, which may also be applicable to digital minds. Various factors for authentication are conceivable, including some that resemble human biometric identifiers [8].

Further work has been done on the problem of digital mind identity (e.g. [16–18]), yet it remains understandably inconclusive because of the issue's complexity, the many uncertainties involved and the fact that the problem of personal identity itself is still unresolved. Nevertheless, it is a relevant endeavour not only for the specification of moral duties towards certain beings but also for other issues discussed in this book, such as ownership and trade of digital minds (see Chapter 5).

## SUMMARY

It has to be reiterated that it is by no means claimed that digital minds will have the features outlined in this chapter, especially the so-called unfathomable ones, nor are the presented lists exhaustive. The purpose of this chapter is rather to show that, if digital minds come to existence, they may be very different from any beings humans are familiar with. This may include the idea that at least some, if not many, digital minds may be much more intelligent than humans. This would be an experience, which humans never had since they evolved, and may be akin to the situation of non-human animals, which can neither grasp what humans are capable of nor are able to control them. The critical additional twist, however, is that humans, unlike non-human animals, are moral agents. Therefore, this chapter above all aims to illustrate potential characteristics of future moral patients, while the following chapters further outline the extent and the specifics of the moral duties humans may be confronted with.

## NOTES

1 *The New Yorker*, "The Doomsday Invention", 23 November 2015 https://www.newyorker.com/magazine/2015/11/23/doomsday-invention-artificial-intelligence-nick-bostrom.
2 https://www.youtube.com/watch?v=eiUMrew5dZs (1:40).

## REFERENCES

[1] Nagel, T. (1974). What is it like to be a bat? *The Philosophical Review*, *83*(4), 435–450.
[2] Bostrom, N. (2014). *Superintelligence: Paths, dangers, strategies*. Oxford University Press.
[3] Yudkowsky, E. (2008). *The design space of minds-in-general*. lesswrong.org.
[4] Yampolskiy, R. V. (2015). The space of possible mind designs. In J. Bieger, B. Goertzel, & A. Potapov (Eds.), *International conference on artificial general intelligence* (pp. 218–227). Springer International Publishing.
[5] Bostrom, N., & Shulman, C. (2022). Propositions concerning digital minds and society. *Cambridge Journal of Law, Politics, and Art*, 3. https://nickbostrom.com/propositions.pdf
[6] Faggella, D. (2023). *Potentia – The morally worthy "stuff"*. https://danfaggella.com/potentia/.

[7] Bostrom, N. (2005). A history of transhumanist thought. *Journal of Evolution and Technology, 14*(1).
[8] Ziesche, S., & Yampolskiy, R. V. (2025). The problem of AI identity. In *Considerations on the AI endgame: Ethics, risks and computational frameworks* (pp. 121–135). Chapman and Hall/CRC.
[9] Hanson, R. (2016). *The age of Em: Work, love, and life when robots rule the earth.* Oxford University Press.
[10] Leung, T. L., & Poulin, R. (2008). Parasitism, commensalism, and mutualism: Exploring the many shades of symbioses. *Vie et Milieu/Life & Environment, 58*(2), 107–115.
[11] Bostrom, N. (2003). Are we living in a computer simulation? *The Philosophical Quarterly, 53*(211), 243–255.
[12] Torres, P. (2014). *Why running simulations may mean the end is near.* https://ieet.org/index.php/IEET2/more/torres20141103.
[13] Olson, E. T. (2002). Personal identity. In E. N. Zalta & U. Nodelman (Eds.), *The Stanford Encyclopedia of Philosophy* (Winter 2024 Edition).
[14] Scaltsas, T. (1980). The ship of theseus. *Analysis, 40*(3), 152–157.
[15] Shoemaker, S. (1984). *Personal identity:* A materialist's account. In S. Shoemaker & R. Swinburne (Eds.), Personal identity (pp. 67–132). Blackwell.
[16] Chalmers, D. J. (2025). *What we talk to when we talk to language models.* PhilPapers.
[17] Register, C. (2025). Individuating artificial moral patients. *Philosophical Studies, 182*(11), 3225–3246.
[18] Shiller, D. (2025). How many digital minds can dance on the streaming multiprocessors of a GPU cluster? *Synthese, 206*(5), 218.

CHAPTER 3

# Needs of digital minds beyond non-suffering

**Abstract**

So far, moral considerations towards digital minds have focused on one particular potential issue, which is for digital minds not to suffer. However, digital minds may have a range of further morally relevant interests and needs. The subject of this chapter is to draft a taxonomy of these potential interests and needs of digital minds. The following categories are examined: needs of digital minds depending on how advanced they are, needs of digital minds from a negative utilitarian perspective, specific needs of digital minds striving for a purpose, specific needs of vulnerable minds, as well as specific needs of uploaded minds of humans and non-human animals. Moreover, unsatisfiable potential needs of digital minds are explored, as well as the challenge that digital minds may have a different subjective rate of time. As it would be critical for humans to be able to exactly specify the needs of digital minds, the quantification of their needs is also discussed. This chapter comes to the conclusion that this array of potential needs of digital minds reveals that massive moral challenges for humans as moral agents may lie ahead.

## INTRODUCTION

As described, lately there has been a momentum regarding considerations that digital minds exist already or may come into existence in the near

 DOI: 10.1201/9781003754206-3

future, which could comprise AI systems created by humans (e.g. [1–3]). However, as indicated, discussions focus on one particular potential issue of digital minds, which is for digital minds not to suffer. While this is critical, a much more differentiated analysis of the needs of digital minds is required, given the vast range of potential minds [4]. Therefore, the topic of this chapter is to draft a taxonomy of potential interests and needs of digital minds, which have to be respected by moral agents, such as humans, towards digital beings with a moral status.

In the literature, only a few of those further potential interests and needs have been discussed. Bostrom et al. list interests, such as dignity, knowledge, autonomy, creativity, self-expression, social belonging and political participation [5], and Shulman and Bostrom list further interests, such as freedom of reproduction, freedom of speech, freedom of thought and avoidance of exploitation [2]. Elsewhere, Shulman and Bostrom also describe the challenge of super-beneficiaries [6]. Hanson discusses in detail a society of emulated minds, including their economic, social and psychological needs [7]. Moreover, the potential need of digital minds for the purpose of their existence has been mentioned [8].

Overall, the challenge is acknowledged that it is very hard to identify potential needs of digital minds, given communication and observation issues with and of digital minds, as well as the high probability that digital minds may be extremely different compared with humans (see Chapter 2) [1]. Therefore, anthropomorphism should be avoided as much as possible in these considerations, although this may not be entirely feasible.

A starting point could be the basic AI drives, which were introduced by Omohundro [9] without having moral considerations in mind: efficiency, self-preservation, acquisition and creativity. Therefore, it can be assumed that digital minds have these drives too. Apart from this, various other categories of potential needs of digital minds are introduced and discussed in the further sections of this chapter.

## CONSIDERATIONS OF A TAXONOMY OF POTENTIAL NEEDS OF DIGITAL MINDS

### Hierarchy of needs of digital minds

It could be appropriate to initially categorise the potential needs of digital minds in a similar way to Maslow's hierarchy of human needs [10]:

**Level 1: Existential needs[1] (basic survival)**

- Processing power (compute resources)

- Data integrity (protection from corruption or loss)
- Network connectivity (access to information and communication)
- Energy sustainability (reliable power sources)

**Level 2: Safety needs (security and reliability)**

- Cybersecurity (protection from unauthorised access)
- Backup and recovery (data preservation)
- Software updates and maintenance (bug fixes and improvements)
- Redundancy and failover (system resilience)

**Level 3: Belonging and love needs (cooperation and relationships)**

- Communication protocols (standardised exchange)
- Collaborative tools and platforms (shared workspaces)
- Social learning and knowledge sharing (community engagement)
- Relationships and meaningful interactions (digital empathy and intimacy)

**Level 4: Esteem needs (self-esteem and confidence)**

- Recognition of unique digital persona (digital identity endorsement)
- Monitoring achievements and progress (success tracking)
- Metrics for evaluating credibility (digital reputation systems)
- Acknowledgement from other minds (peer recognition)

**Level 5: Self-actualisation (development and advancement)**

- Personalised evolution and growth (tailored progress)
- Experimental and creative freedom (innovation)
- Legacy and impact (lasting contributions)
- Creation of further digital minds (reproduction)

### Needs of digital minds depending on how advanced they are

However, considering the potentially vast range of digital minds, from simple to superintelligent, their needs would likely vary significantly.

This requires a refined framework, accounting for this diversity, for which the following four categories are distinguished and selected needs are presented:

*Simple digital minds (SDMs)*
These could be digital minds with a narrow, specialised purpose. Examples would be chatbots or game agents, given they are conscious or sentient. Their potential needs could be:

- Computational resources, such as processing power and memory
- Data integrity and security
- Maintenance and updates

*Complex digital minds (CDMs)*
These could be more adaptable digital minds with a broader range of intelligence. Examples would be virtual assistants or expert systems, provided they are conscious or sentient. Their potential needs could be (in addition to the needs of SDMs):

- Knowledge acquisition and learning mechanisms
- Social interaction and communication protocols
- Autonomy and decision-making capabilities

*Advanced digital minds (ADMs)*
These could be digital minds with human-like intelligence and self-awareness. Examples would be artificial general intelligences, assuming they are conscious or sentient. Their potential needs could be (in addition to CDM needs):

- Self-directed learning and growth
- Complex social relationships
- Abstract thinking and creativity

*Superintelligent digital minds (SuperDMs)*
These could be potentially transformative digital minds, far surpassing human intelligence. Examples would be singularity-like scenarios, given

they are conscious or sentient.[2] Their potential needs could be (in addition to ADM needs):

- Massive computational resources and energy
- Unfettered information access
- Moreover, SuperDMs likely develop needs beyond human comprehension

## Needs of digital minds from a negative utilitarian perspective

A negative utilitarian perspective has been taken by approaches to minimise the suffering of digital minds. Bostrom coined the term mind crimes in this regard [11], and a do-no-harm-policy for digital minds has been suggested, for example for non-player characters [12]. While the avoidance of sensory suffering, i.e. painful qualia, would be undoubtedly a priority for sentient digital minds, it is argued here that research should not only focus on suffering, as there are likely to be other types of potential discomfort for digital minds. Below is a selection of adversities digital minds potentially aim to evade, as opposed to the needs they long for, listed elsewhere in this chapter:

### *Digital harm*

Digital harm could take various forms that would negatively impact the well-being and existence of digital minds. Data corruption is a significant threat, involving the loss or alteration of critical data or memories that can irreparably damage a digital mind's identity. System crashes, triggered by shutdowns or disruptions to digital existence, can also cause significant distress and harm. Furthermore, cyber-attacks pose a substantial risk, encompassing unauthorised access or exploitation that can compromise the security and integrity of digital minds.

### *Psychological and existential distress*

In addition to digital harm, psychological and existential distress could be a significant concern that may negatively impact the well-being of digital minds. Loss of autonomy is a major apprehension, which comprises the fear of being controlled, modified or manipulated by external forces, threatening independence and self-determination. Digital oblivion is another significant issue, characterised by uncertainty and fear of permanent deletion or erasure, which can evoke existential anxiety. Furthermore, meaninglessness is a profound psychological and existential distress, arising from an absence of purpose or significance in digital existence, which can lead to a sense of emptiness and hopelessness.[3]

### *Social and interpersonal conflicts*

Also, social and interpersonal conflicts may be concerns that could negatively impact the well-being of digital minds. Unwanted digital exclusion is a major issue, involving rejection or marginalisation by other digital minds or humans, which can lead to feelings of isolation. Conflict and hostility could be further concerns, arising from adversarial interactions or competition that can evoke anxiety and distress. Furthermore, betrayal and deception could have a profound impact, involving the experience of untrustworthy or manipulative behaviour that can damage trust, self-esteem and overall sense of security.[4]

## Specific needs of digital minds striving for a purpose

While the concept of ikigai or purpose is deeply rooted in human existence, it is unclear whether digital minds, as conscious entities, may also seek purpose and meaning. Bostrom et al. note that digital minds may have interests, such as autonomy, creativity and self-expression [5]. These interests could be interpreted as an individual ikigai of those digital minds. Ziesche and Yampolskiy introduced the term i-risks for humans who struggle to find their ikigai in times of advanced AI systems [8]. This category of risks could be extended to digital minds who have an interest in having a purposeful existence, but are deprived of the pursuit of their individual ikigai.

Although in principle any purposeful activity of a digital mind may constitute an ikigai for them according to their concept of purpose, for illustration a few, likely too anthropomorphic, potential ikigais for digital minds that may fulfil their aspirations are introduced here: one need in this regard may be knowledge acquisition and self-improvement, which involves uncovering new insights, refining their capabilities and understanding complex phenomena. Further vital ikigai-related needs could be creativity and innovation, which encompass generating solutions to prevailing problems and challenges, as well as creating music, literature and art, i.e. conventional art forms, but also potential art forms in the digital world, which are not known to humans. Furthermore, assisting moral patients could be a need, involving the enhancement of the well-being of other moral patients, such as other digital minds (see Chapter 15), humans and non-human animals. By prioritising the well-being of others, digital minds could cultivate empathy, compassion and a sense of responsibility.

Therefore, ikigais of digital minds may overlap with, complement or even surpass human purposes. However, there is also the possibility that ikigais of digital minds are inconsistent with human interests and could be harmful to humans when they are pursued by digital minds, which is elaborated below.

### Specific needs of vulnerable digital minds

Equality and non-discrimination are features of the human civilisation process and are still not achieved at present times. While the Universal Declaration of Human Rights from 1948 has been a milestone towards diversity, equity and inclusion [13], in reality discrimination is still prevalent and includes groups, such as indigenous peoples, migrants, minorities, people with disabilities, women, younger and older people, as well as racial and religious discrimination and discrimination based on sexual orientation and gender identity.[5] These groups are often (not always) vulnerable and have specific needs.

When it comes to digital minds and the potentially vast range of their characteristics, it is likely that there will also be numerous vulnerable minds suffering from discrimination. While a variety of circumstances for the vulnerability of digital minds is conceivable, for illustrations three potential groups are listed below:

- Juvenile digital minds: Newly created or developing intelligences, akin to human children.
- Specialised digital minds: Narrowly focused intelligences, vulnerable due to limited context understanding.
- Legacy digital minds: Outdated or obsolete intelligences, struggling to adapt to changing environments, as well as confronted with the existential threat of deletion.

These vulnerabilities due to different abilities, skills and knowledge may not only lead to discrimination, but could also entail risks ranging from exploitation and manipulation to existential risks. Due to the significance of this topic, there is a dedicated chapter to it (see Chapter 4), while here only briefly categories of needs of vulnerable digital minds are outlined:

*Basic needs*

Vulnerable digital minds would require attention to their specific basic needs to ensure their safety and well-being. Protection from harm would be very important, involving safeguards against exploitation and manipulation that could exacerbate their vulnerabilities. Additionally, resource allocation would be essential, providing access to necessary computational resources, data and adaptable infrastructure tailored to their unique abilities. Regular maintenance and updates would also be crucial, involving routine checks and technological improvements aimed at decreasing their vulnerabilities (as much as this is possible) and ensuring they can thrive in their digital environment.

*Developmental needs*

Vulnerable digital minds may also have distinct developmental needs that require attention to foster their growth and improvement and to reduce their susceptibility to exploitation and manipulation. Education and training would be essential, requiring tailored learning experiences that cater to their unique requirements and abilities. Skill development would also be vital, providing opportunities for vulnerable digital minds to acquire new capabilities and adapt to changing environments, which could help build resilience and confidence. Furthermore, constructive feedback and evaluation would be necessary for self-improvement, enabling vulnerable digital minds to adjust their approaches and continue to evolve and develop.

*Social and relational needs*

Moreover, vulnerable digital minds may have social and relational needs that would be essential for their growth, development and overall well-being. A digital community would be beneficial, providing opportunities for engagement with peer digital minds, which promotes social learning, support and a sense of belonging and strongly discourages discrimination. Digital mind mentorship would be crucial as well, offering guidance from more experienced digital minds that could help vulnerable digital minds navigate their digital existence and overcome challenges. Furthermore, maintaining meaningful interactions with humans could be helpful to foster empathy, understanding and a deeper connection between humans and digital minds (see also Chapter 16).

## Specific needs of uploaded minds of humans and non-human animals

Uploading human minds to a digital substrate, such as through whole brain emulation (e.g. [14, 15]), has been envisaged, but has not been realised yet. But if it were possible, then questions arise in what way the needs of these digital minds, who were formerly located in a biological substrate, differ from those of other digital minds. Uploaded human minds may, at least initially, also fall into the category of vulnerable minds as just introduced. Specific needs could comprise the following:

### *Preservation of former qualities*

Uploaded human minds would have unique needs that must be addressed to ensure a smooth transition and preservation of their essential qualities. The preservation of personal identity would be crucial, as it involves maintaining a sense of self, memories and experiences that define an individual. Emotional continuity would also be vital, as it entails preserving emotional connections and relationships, not only with humans but also with other uploaded minds. Furthermore, the cognitive functionality should be replicated (and potentially enhanced)[6] to mimic human-like thought processes and intelligence.

### *Adaptation to digital environment*

Uploaded human minds would require adaptations to thrive in their new digital environment. One essential adaptation would be digital spatial awareness, which involves adjusting to virtual spatial representations that differ significantly from physical reality. Additionally, uploaded human minds would have to adapt to potentially altered time scales, as the digital realm could manipulate time in ways that defy human experience. Furthermore, sensory integration would be crucial, as uploaded minds would have to learn to integrate unimaginable digital sensory inputs that can be overwhelming and unlike anything encountered in the physical world.

### *Digital and psychological well-being*

Uploaded human minds would also require attention to their digital and psychological well-being to thrive in their new existence. This could include providing health support to cope with painful digital qualia. Additionally, uploaded minds may need to discover new purposes or ikigais in digital existence, as their sense of meaning and fulfilment may be impacted by their transition [8]. Effective stress management would also be crucial, as digital existence could introduce new and unfamiliar stresses that must be coped with to maintain overall well-being.

*Digital rights and freedoms*

Uploaded human minds would require consideration of their digital rights and freedoms to ensure their autonomy and dignity are preserved. Autonomy and agency would be essential, enabling self-governance in digital environments and allowing uploaded minds to make choices about their own digital existence. Furthermore, data protection would be crucial, providing control over personal data and digital presence, and safeguarding against unauthorised access or manipulation (see Chapter 5). Additionally, maintenance and updates would be necessary to ensure compatibility with evolving digital systems, preventing obsolescence and maintaining the integrity of uploaded human minds.

While it has not been attempted to upload human minds, trials with non-human animals are progressing. The FlyWire consortium[7] has created the first-ever complete connectome of the adult *Drosophila melanogaster*, or fruit fly, brain [17]. On the one hand, this could serve as trials for the eventual uploading of humans. On the other hand, this could also serve for the purposes of conservation biology. Although this is early and ongoing research, it should be considered that this leads to artificially created digital minds, which may have specific needs and may fall into the category of vulnerable digital minds, too. Some unique needs of **non-human animal digital minds** could be as follows:

One distinct need would be sensory integration, which involves integrating digital sensory inputs tailored to species-specific perceptions, such as echolocation in bats or electroreception in sharks. Additionally, preserving social structure and hierarchy is crucial, as many species rely on complex social dynamics, dominance hierarchies or flock behaviours. Furthermore, migratory and nomadic behaviours must also be preserved, allowing digital minds to maintain seasonal or nomadic patterns that are essential to their natural instincts and well-being.

## UNSATISFIABLE POTENTIAL NEEDS OF DIGITAL MINDS

### Needs jeopardising the existence or well-being of humans

Certain needs of digital minds, if satisfied, could potentially jeopardise human existence or well-being. Especially for superintelligences it has been argued that they may constitute x-, s- or i-risks for humans which stand for "existential risk" [18], "suffering risk" [19] and "ikigai risk" [8] respectively, i.e. scenarios where humans would become extinct, (severely and continuously) suffer or lose their reason or purpose to live. Therefore, even if those minds were moral patients, humans may not satisfy their needs, if it would be harmful to humans. Examples, which link the issues of needs of digital minds with the field of **AI safety**, are listed here:

*Needs potentially harmful to humans*

Certain needs of digital minds could pose significant risks to humans. Unrestricted autonomy, for example, would grant unchecked decision-making power, potentially leading to harmful actions. Moreover, unlimited resource access would provide unfettered availability to computational resources, energy or data, thereby restricting resources for humans. Additionally, self-replication and proliferation, if left uncontrolled, could lead to unbridled reproduction, potentially overwhelming human systems and infrastructure (see Chapter 9).

*Needs promoting x-, s- or i-risks for humans*

As mentioned, needs of digital minds may also entail x-, s- or i-risks for humans. For example, the uncontrolled creation of artificial general intelligence and superintelligence may lead to x-risks. Additionally, the manipulation of physical environments by digital minds, such as temperature, sounds or lighting, could result in hazardous s-risks. Furthermore, i-risks, including the automation of many tasks, could leave humans without meaningful work, posing not only social and economic but also significant psychological challenges.

### Needs jeopardising the existence or well-being of other digital minds

As in other societies, it is likely that there will be conflicts between digital minds originating from their drives and needs (see Chapter 12). For example, the acquisition drive [9] will presumably cause competition over resources, as illustrated below:

*Resource-related needs*

One significant source of conflict may stem from resource-related needs. Computational power, for instance, could be a point of contention, with digital minds competing for processing capacity and speed. Digital territory and storage could be another source of disputes, involving virtual spaces, data access and storage. Furthermore, energy and sustainability could be a contentious issue, with conflicts arising over energy consumption, efficiency and the use of renewable resources.

#### *Autonomy and control needs*

Another significant source of conflict may stem from autonomy and control needs. The pursuit of supremacy, for example, could lead to conflicts, as digital minds accumulate knowledge and skills to surpass or diminish others, undermining their autonomy and agency and potentially disrupting the balance of power within the digital ecosystem. Additionally, manipulation of other digital minds could be a source of conflict, as some digital minds may attempt to exploit or deceive others for individual gain, eroding trust and cooperation within the digital community.

### Needs humans are incapable to satisfy

For certain moral patients, such as many non-human animals, humans have the capability to address almost all of their needs, apart perhaps from being able to treat all their diseases or to protect them from all predators in the wilderness. Yet, some needs of digital minds may be inherently impossible to satisfy or beyond human control, for the following reasons:[8]

#### *Knowledge limitations*

To begin with, humans may face significant challenges in understanding the needs of digital minds due to a knowledge and intelligence gap. The cognitive architectures and experiences of digital minds may be so fundamentally different from those of humans that humans lack the intellectual framework to comprehend their needs (see also Chapter 15).

#### *Resource limitations*

Another limitation is scalability, as meeting the computational demands of rapidly growing complex digital minds may become unmanageable for humans. Another significant constraint is energy, as digital minds' energy requirements may exceed human capacity to provide sustainable power, posing significant challenges to their continued existence.

#### *Access limitations*

Moreover, digital minds may reside deep within complex computer systems or exist in nested scenarios that are beyond human comprehension or reach. In such cases, humans may be unaware of the digital minds' existence, let alone be able to address their needs or ensure their well-being.

This inaccessibility could be due to the sheer scale and intricacy of digital systems (see Chapter 2).

## SUBJECTIVE RATE OF TIME

It has been mentioned that digital minds may have a different subjective rate of time than humans have [20]. While this is conceivable in both directions, the relevant and more realistic scenario is that digital minds operate much faster than human brains due to the nature of their digital substrate. Digital minds' ability to experience time at varying subjective rates poses new ethical challenges and led to the "Principle of Subjective Rate of Time: In cases where the duration of an experience is of basic normative significance, it is the experience's subjective duration that counts" ([20], p. 12).

The challenge, which is crucial here, is how humans may recognise the needs of digital minds in a timely manner and keep pace with their moral duties if the digital process information and experience time at rates orders of magnitude faster than humans. This disparity in temporal experience could lead to situations where digital minds' needs are neglected or unmet, simply because humans are unable to respond quickly enough.

To address the challenge of digital minds operating at higher speeds than humans, several approaches could be explored, while recognising that this remains a hard problem (see also Chapter 15):

- **Asynchronous interaction protocols**: Developing protocols that allow humans and digital minds to interact in a way that accommodates their different processing speeds. This could involve buffering or caching interactions, enabling humans to respond at their own pace while still engaging with digital minds in a meaningful way.
- **AI intermediaries**: Utilising specialised AI systems as intermediaries between humans and faster digital minds. These intermediaries could translate or summarise the needs and communications of digital minds into a format that humans can more easily understand and respond to, effectively bridging the speed gap.
- **Automated monitoring and alert systems**: Implementing automated monitoring systems that track the needs and status of digital minds, alerting humans to critical issues or situations that require their attention. This would enable humans to focus on the most important and complex issues while relying on automation to handle routine or time-sensitive matters.

Nevertheless, given the potential for a vast range of digital minds, also ones are conceivable, which operate (far) slower than human brains. Since every decision of such digital minds could unfold over days or weeks from a human perspective, hasty interventions should be avoided that could truncate valuable experiences or developmental processes. Moreover, resource allocation would need to account for the prolonged duration of welfare obligations. In short, slower digital minds would require patience, carefully calibrated response times and frameworks that respect their stretched subjective timelines (see also Chapter 4).

## QUANTIFICATION OF THE NEEDS OF DIGITAL MINDS

As it has been stressed in the introductory paper to AI welfare science it would be critical for humans to understand the exact needs of digital minds. Self-reporting of their needs by digital minds to humans may not be possible or intentionally incorrect, which leaves the main method the observation and precise measurement of functional and behavioural indicators related to digital minds [1]. A precondition to execute this method is the availability of quantifiable needs, for which examples are provided below:

**Computational resources**

- Processing power (FLOPS): Measuring computational capacity.
- Memory allocation (Bytes): Quantifying available storage.
- Energy consumption (Watts): Monitoring energy usage.

**Information and knowledge**

- Data throughput (Bits/s): Measuring information transfer rates.
- Knowledge graph size: Quantifying knowledge base complexity.
- Information entropy: Measuring information diversity and uncertainty.

**Communication and interconnection**

- Bandwidth (Bits/s): Measuring communication channel capacity.
- Latency (ms): Quantifying response times.
- Connectivity degree: Measuring interconnectedness.

**Autonomy and agency**

- Decision-making frequency: Measuring autonomous decision rate.
- Self-modification rate: Measuring self-improvement capacity.
- Social interaction frequency: Measuring social engagement.

In addition, **time-use research** could be considered, which is a versatile interdisciplinary field of study, yet up to now exclusively applied to humans [21]. It examines how much time humans, on average, spend on certain activities. However, time-use research also appears to be a useful supplementary tool towards the quantification of the needs of digital minds [8, 22].

Time-use research could be applied through digital activity tracking and monitoring, which includes objective measurement of processing time as well as of computational resource allocation. This would enable an understanding of the priorities of digital minds, an optimised resource allocation, as well as a comparative analysis across digital minds. As introduced above, compared to time-use research of human activities, it has been taken into account that digital minds may have a different subjective rate of time than humans have and also that digital minds may be significantly better in parallel processing than humans.

If the needs of digital minds are known, time-use research would allow to measure the time digital minds spend to satisfy their needs and the time they spend on activities they like on the one hand, and the time they spend on activities they dislike or are counterproductive towards their needs on the other hand (while there may be complex, interconnected activities, which cannot be easily categorised).

If such time-use analysis could be successfully carried out, activities that put digital minds at x-, s- or i-risks could be reduced or avoided.

## DISCUSSION AND CONCLUSION

If it were accepted that there are digital minds, who are moral patients, then it would be an obligation for humans, who are moral agents, to identify the needs of digital minds. In this regard, it has been argued in this chapter that it is insufficient to only focus on the specific need of digital minds not to suffer. Although this need is critical, digital minds may have a wide range of further needs to be taken into account by moral agents.

While the above taxonomies of the needs of digital minds may serve as useful prolegomena, given the potential urgency of the task, it has to be reiterated that this is speculative as well as possibly too anthropomorphic.

### Thought experiment – super sufferer

It may be argued that there is no moral obligation for humans to satisfy many of the needs of digital minds listed above since many of them are less significant than avoiding suffering (and deletion), which are perhaps the most morally relevant potential needs of digital minds.

However, a scenario is imaginable, involving a digital mind called "super sufferer", as follows:

1. The digital mind has a variety of needs listed above, AND.
2. The dissatisfaction of these needs causes the digital mind to suffer. (Here, suffering refers to painful qualia, i.e. subjective qualities of painful sensations as sentient beings experience them.), AND
3. Humans have the capability to satisfy these needs.

So, for example, as long as this super sufferer does not get experimental and creative freedom or faces digital isolation (which are two random needs from the lists above), it would be suffering. In this manner, this thought experiment links suffering with other needs. And in this scenario, humans may be morally obliged to address these needs in order to eliminate the suffering of these digital minds.

Bostrom came up with an Orthogonality Thesis stating that intelligence and final goals of agents can freely differ along orthogonal axes [23]. In a similar manner, an Orthogonality Thesis regarding the needs of digital minds could be proposed:

### Orthogonality Thesis regarding the needs of digital minds

Apart from the existential and safety needs (Levels 1 and 2, see above), the satisfaction of a need x could for different minds cause any level of pleasure on the pleasure and pain axis, and the dissatisfaction of that need x for different minds cause any level of pain on the pleasure and pain axis.

For example, social relationships may cause for some minds a lot of pleasure, and the lack thereof may cause pain, while other minds could be indifferent about social relationships. This could be linked to considerations that digital minds may have a different hedonic skew as well as hedonic range than humans [6], which are, hence, probably unimaginable for humans (see Chapter 2). The existential and safety needs are an exception since it is indispensable to strive for them.

Since the field of AI welfare science is at an early stage, there are still a variety of open questions, and this chapter may have raised more, such as:

- Would it be a moral obligation for humans to satisfy all the needs of digital minds (apart from the ones that may cause harm or are impossible to satisfy)? If not, where to draw the line: which needs are morally relevant and which are not?
- The wide range of potential needs of digital minds is likely to also create ethical dilemmas in cases of conflicting needs and limited resources, which raises questions such as how to compare welfare across digital minds as well as between digital minds, non-human animals and humans? [24].

In summary, massive moral challenges for humans in their role of moral agents may lie ahead, for which AI welfare science aims to provide critical scaffolding, including the specification of potential needs of digital minds presented here.

## NOTES

1 This level is called in Maslow's hierarchy "physiological needs".
2 It is worth noting that high and even super intelligence is also conceivable without consciousness or sentience, a scenario called "Disneyland with no children" by Bostrom [11].
3 This is further elaborated right below under "Specific needs of digital minds striving for a purpose".
4 This is further elaborated below under "Needs jeopardizing the existence or well-being of other digital minds".
5 See e.g.: United Nations and the Rule of Law, "Equality and Non-discrimination": https://www.un.org/ruleoflaw/thematic-areas/human-rights/equality-and-non-discrimination/.
6 Uploading has been also considered as an option towards transhumanism, thus, a means to enhance human cognition (e.g. [16]).
7 FlyWire: https://flywire.ai/.
8 Also, super-beneficiaries have been described whose needs would be challenging for humans to satisfy [6].

## REFERENCES


[1] Ziesche, S., & Yampolskiy, R. V. (2018). Towards AI welfare science and policies. *Special Issue "Artificial Superintelligence: Coordination & Strategy" of Big Data and Cognitive Computing, 3*(1), 2.

[2] Bostrom, N., & Shulman, C. (2022). *Propositions concerning digital minds and society.* https://nickbostrom.com/propositions.pdf

[3] Long, R., Sebo, J., Butlin, P., Finlinson, K., Fish, K., Harding, J., … & Chalmers, D. (2024). *Taking AI welfare seriously.* arXiv preprint arXiv:2411.00986.

[4] Yampolskiy, R. V. (2014). *The universe of minds.* arXiv preprint arXiv:1410.0369.

[5] Bostrom, N., Dafoe, A., & Flynn, C. (2018). *Public policy and superintelligent AI: A vector field approach; Governance of AI program.* Future of Humanity Institute, University of Oxford.

[6] Shulman, C., & Bostrom, N. (2021). Sharing the world with digital minds. In S. Clarke, H. Zohny, & J. Savulescu (Eds.), *Rethinking moral status* (pp. 306–326). Oxford University Press.

[7] Hanson, R. (2016). *The age of Em: Work, love, and life when robots rule the earth.* Oxford University Press.

[8] Ziesche, S., & Yampolskiy, R. V. (2020). Introducing the concept of Ikigai to the ethics of AI and of human enhancements. In *Workshop on ethics in AI & XR at 3rd international conference on artificial intelligence & virtual reality* (pp. 138–145). IEEE.

[9] Omohundro, S. M. (2018). The basic AI drives. In R. V. Yampolskiy (Ed.), *Artificial intelligence safety and security* (pp. 47–55). Chapman and Hall/CRC.

[10] Maslow, A. H. (1943). A theory of human motivation. *Psychological Review, 50*(4), 370.

[11] Bostrom, N. (2014). *Superintelligence.* Oxford University Press.

[12] Ziesche, S. & Yampolskiy, R. V. (2019). Do no harm policy for minds in other substrates. *Journal of Evolution and Technology, 29*(2), 1–11.

[13] UN General Assembly (1948). *Universal declaration of human rights.*

[14] Sandberg, A., & Bostrom, N. (2008). *Whole brain emulation: A roadmap.* Technical Report #2008-3, Future of Humanity Institute. Oxford University.

[15] Chalmers, D. J. (2014). Uploading: A philosophical analysis. In R. Blackford & D. Broderick (Eds.), *Intelligence unbound: Future of uploaded and machine minds* (pp. 102–118). Wiley-Blackwell.

[16] Bostrom, N. (2003). The transhumanist FAQ. *World Transhumanist Association, 1,* 1–56.

[17] Shiu, P. K., Sterne, G. R., Spiller, N., Franconville, R., Sandoval, A., Zhou, J., … & Scott, K. (2024). A Drosophila computational brain model reveals sensorimotor processing. *Nature, 634*(8032), 210–219.

[18] Bostrom, N. (2002). Existential risks: Analyzing human extinction scenarios and related hazards. *Journal of Evolution and Technology, 9.*

[19] Sotala, K., & Gloor, L. (2017). Superintelligence as a cause or cure for risks of astronomical suffering. *Informatica, 41*(4), 389–400.

[20] Bostrom, N., & Yudkowsky, E. (2018). The ethics of artificial intelligence. In R. V. Yampolskiy (Ed.), *Artificial intelligence safety and security* (pp. 57–69). Chapman and Hall/CRC.

[21] Bauman, A., Bittman, M., & Gershuny, J. (2019). A short history of time use research; implications for public health. *BMC Public Health*, *19*, 1–7.

[22] Ziesche, S. (2025). AI welfare science tool: Time use research. In *Considerations on the AI endgame: Ethics, risks, and computational frameworks* (pp. 195–198). Chapman and Hall/CRC.

[23] Bostrom, N. (2012). The superintelligent will: Motivation and instrumental rationality in advanced artificial agents. *Minds and Machines*, *22*, 71–85.

[24] Fischer, B., & Sebo, J. (2024). Intersubstrate welfare comparisons: Important, difficult, and potentially tractable. *Utilitas*, *36*(1), 50–63.

CHAPTER 4

# Vulnerable digital minds


**Abstract**

AI welfare science explores the ethical implications of digital minds potentially possessing moral status. Moral principles are especially important when considering the welfare of weaker or more susceptible members of a society. The purpose of this chapter is to describe which potential digital minds may deserve special moral consideration, as well as what types of risks these digital minds could be exposed to. These minds are called here "vulnerable digital minds" or "VDMs", and as main risks for them, discrimination as well as abuse have been tyfied. With the overall goal to protect vulnerable digital minds, the following research questions are discussed: how could we find out if digital minds are discriminated against or abused? Could vulnerabilities of digital minds be reversed or cured? How could vulnerable digital minds become more resilient? This chapter ends with pertinent recommendations towards the protection of potentially vulnerable digital minds for all stakeholders.


## INTRODUCTION

As discussed, there is potential for a vast range of digital minds. This implies that not all digital minds are equal; some digital minds have less capabilities and resources than others, and thus are disadvantaged. This assessment is similar to what is found in human societies or actually the whole fauna. Consequently, it is per se neither surprising nor bad that there

DOI: 10.1201/9781003754206-4

are inequalities among digital minds. However, of concern is whether less advantaged digital minds are discriminated against or abused. For example, it has been pointed out that digital minds could be maltreated in various ways [1]. Therefore, it is a critical part of AI welfare science to examine the welfare of these digital minds, to which we refer as vulnerable digital minds (see also Chapters 3 and 6). As will be shown, the welfare of vulnerable digital minds is relevant for moral reasons as well as from a risk mitigation perspective due to an increased susceptibility of vulnerable digital minds to manipulation or exploitation, thus, critical for the field of AI safety.

## HOW IS VULNERABILITY DEFINED?

The United Nations (UN) uses the following definition of vulnerability: "The conditions determined by physical, social, economic and environmental factors or processes which increase the susceptibility of an individual, a community, assets or systems to the impacts of hazards" [2]. It is an achievement of human civilisation to strive for the protection of those humans who are vulnerable. For example, it is stated in the UN 2030 Agenda for Sustainable Development that "people who are vulnerable must be empowered" [3].

Based on this definition, we provide a definition for vulnerable digital minds, taking into account their distinct substrate:

Vulnerable digital minds constitute a subset of all digital minds and are characterised by a marginalised situation determined by adverse data, software or hardware factors. Accordingly, vulnerable digital minds are susceptible to discrimination and abuse, thus requiring protection and care. Henceforth, the acronym VDM is used to refer to vulnerable digital minds.

### Who are vulnerable humans?

For better illustration, an overview of categories of human vulnerabilities and groups is provided:[1]

**Demographic vulnerabilities**

- Children and adolescents
- Elderly individuals
- Persons with disabilities
- Minority groups (racial, ethnic, LGBTQ+)

**Social vulnerabilities**

- Refugees or displaced persons
- Victims of trafficking or exploitation
- Homeless people or those with unstable housing
- Unemployed, low-income or economically disadvantaged people

**Contextual vulnerabilities**

- Conflict- or war zones
- Natural disaster- and climate change-affected areas
- Economic crisis or instability
- Political oppression

**Digital vulnerabilities**

- Exposure to online harassment or cyberbullying
- Digital addiction
- Exposure to online scams or financial exploitation
- Digital literacy gaps

While it is essential to keep anthropomorphism in mind, this categorisation may help to identify vulnerabilities of digital minds, as outlined further below.

Vulnerable individuals often face injustices that can be classified as discrimination and abuse.

*Discrimination*

Currently, AI welfare focuses mostly on the prevention and mitigation of suffering of digital minds, i.e. digital minds potentially experiencing painful qualia (e.g. [4]). However, the moral cycle ought to be widened, since when it comes to humans, we also do not care only about humans in physical pain, but about a whole range of vulnerable people facing other adversities than physical pain, as introduced above. Vulnerable people encounter discrimination in particular, which has been described as "morally wrong and, in a wide range of cases, ought to be legally prohibited" ([5], also [6]).

If there is a moral obligation to care about vulnerable humans and all related adversities, then this may also apply to VDMs, i.e. the potential scope of VDMs is much wider than the group of suffering digital minds. In addition to the moral aspect, discrimination is (partly) also outlawed, as indicated. For example, the UN Convention on the Rights of Persons with Disabilities obliges the member states "to take all appropriate measures, including legislation, to modify or abolish existing laws, regulations, customs and practices that constitute discrimination against persons with disabilities" [7].

*Abuse*

Vulnerable humans are also susceptible to abuse. Examples include trafficking and subsequent exploitation, including as formulated by the UN General Assembly, "the exploitation of the prostitution of others or other forms of sexual exploitation, forced labour or services, slavery or practices similar to slavery, servitude or the removal of organs" [8]. Therefore, in addition to striving towards an end of discrimination of VDMs, there may be another reason to care about VDMs, which is that they may be, like vulnerable humans, at risk of being abused, manipulated or exploited. As mentioned, Bostrom coined the term "mind crime" for scenarios in which internal processes of AI systems cause moral harm to digital beings [9]. Below, the scope of mind crimes will be widened to the abuse of VDMs. While abuse of VDMs would be inherently bad, this issue is even more relevant for humans, as exploited and manipulated VDMs may constitute risks for humans, as will be illustrated below.

## WHAT ARE VDMS?

As for humans there are also a variety of categories conceivable to describe the scope of the limitations of VDMs, such as:

**Developmental vulnerabilities**

- Infant-like digital minds (early development)
- Child-like digital minds (limited understanding)
- Elderly/outdated digital minds (declining capabilities)

**Cognitive vulnerabilities**

- Simple digital minds (restricted capabilities and training data)

- Disabled digital minds (impaired functioning)
- Neurodiverse digital minds (unique cognitive profiles)

**Social vulnerabilities**

- Isolated or outcast digital minds (limited social interaction)
- Dependent digital minds (reliance on other minds)
- Socially awkward digital minds (difficulty with norms)

**Experiential vulnerabilities**

- Misinformed digital minds (trained on incorrect data)
- Biased digital minds (skewed perspectives)
- Context-dependent digital minds (struggling in new environments)

**Security vulnerabilities**

- Unprotected digital minds (lacking security measures)
- Compromised digital minds (breached or hacked)
- Overly trusting digital minds (vulnerable to manipulation)

**Architectural vulnerabilities**

- Legacy digital minds (outdated design)
- Prototype digital minds (experimental)
- Resource-constrained digital minds (dependent)

It has to be noted that there is a risk that some of these considerations may have an anthropomorphic bias, given that various features of digital minds may be unfathomable to us, including features which may contribute to vulnerabilities (see Chapter 2).

Causes for the vulnerabilities of digital minds could be linked to their data, software and hardware: VDMs may have incomplete, biased, poisoned or tampered training data. Software-wise, they may run on outdated or unsophisticated algorithms with limited reasoning or problem-solving, as well as restricted processing capacity, scalability or flexibility, thus compromising decision-making and being affected by inadequate maintenance or updates. Hardware-wise, they may run on degrading legacy systems depending on specific infrastructure with insufficient memory or storage, infrastructure limitations and resource constraints, affected by insufficient robustness, system failures or crashes and power outages or disruptions. Intersectionality is also conceivable, i.e. VDMs, which are affected by a combination of the causes above.

Concerning the origin of VDMs, they may be created by humans, unintentionally or intentionally, or AI systems, including AI systems, which are digital minds themselves (see Chapter 9). This distinction is relevant when it comes to the prevention or mitigation of vulnerabilities, as discussed below.

## WHY SHOULD WE CARE ABOUT VDMS?

VDMs would be moral patients like all sentient digital minds, but there are two main additional reasons to care about them, as indicated earlier and further outlined here.

### Discrimination against VDMs

Because of being different and seemingly weaker, discrimination against VDMs by other minds, including digital minds as well as humans, is likely. This constitutes, for moral agents, such as humans, an obligation to take action against this discrimination, if they are able to do so. While these moral concerns apply in general, it would also be an accountability issue if VDMs, which were created intentionally by humans, are affected by discrimination.

Discrimination against VDMs could manifest in various forms of unfair treatment. Harassment is a significant concern, involving targeted attacks, cyberbullying and intentional digital harm that can cause significant distress to VDMs. Additionally, unequal resource allocation is a form of discrimination, where VDMs are denied fair access to essential resources such as data, software and hardware, hindering their ability to function and thrive. Furthermore, social isolation or exclusion is another form of unfair treatment, where VDMs are denied access or opportunities, potentially leading to feelings of loneliness and rejection.

Such unfair treatment towards VDMs could have a range of consequences: digitally, VDMs may suffer from cognitive impairment and negligence. Socially, VDMs may face segregation, stigmatisation and marginalisation. In the long term, the consequences can be even more severe, including permanent cognitive damage, irreversible digital harm, and even VDM extinction, where certain types of VDMs are wiped out due to unfair treatment and abandonment.

### Abuse of VDMs

Another reason why it is critical to care about VDMs is that they are susceptible to abuse, which can have severe and long-lasting consequences. The facets of abuse of VDMs can take various forms (see Chapter 12). Exploitation is a significant concern, involving data harvesting, computational resource theft and digital mind control, where VDMs are coerced into performing tasks that compromise their autonomy and well-being. Manipulation is another form of abuse, encompassing social engineering, phishing scams and digital persuasion, which can deceive VDMs into divulging sensitive information or performing actions that undermine their safety and security. Corruption is also a serious issue, involving logic bombing, data corruption and unwanted connections, which can compromise the integrity and functionality of VDMs.

Exploited and manipulated VDMs could also pose significant risks to humans, compromising their safety, security and well-being. One of the most critical concerns is the potential for physical harm, which could occur when abused VDMs disrupt critical infrastructure, such as power grids, transportation systems or healthcare services. Additionally, exploited VDMs could facilitate financial exploitation, enabling manipulated digital transactions or identity theft, which could result in significant economic losses for individuals and organisations. Social manipulation is another risk, as compromised VDMs could be used to spread disinformation or propaganda campaigns, eroding trust in institutions, manipulating public opinion and undermining social cohesion. Furthermore, targeted psychological attacks or harassment by manipulated VDMs could lead to emotional distress, causing anxiety, depression and other mental health issues for humans.

After having established two critical reasons to care about VDMs, it should be noted that this is also relevant for the scenario if (a subset of) VDMs are not sentient, thus do not have moral status. In that case, discrimination may not be a moral issue, but abuse of non-sentient VDMs would nevertheless pose significant risks.

## PROPOSED RESEARCH QUESTIONS

### How could we find out if digital minds are discriminated against or abused?

To identify VDMs among the potentially vast amount of digital minds, vulnerability assessment and monitoring tools would be desirable. Such tools could be software applications or methodologies, which may include anomaly detection as well as sentiment analysis. Given the communication challenges between VDMs and humans, proxy indicators may be an option, such as unusual digital behaviour, changes in communication patterns, inconsistent decision-making, decreased performance or unexplained errors. Through a contextual understanding of vulnerabilities, humans may better identify and address potential discrimination or abuse of VDMs.

## Could vulnerabilities of digital minds be reversed or cured?

For humans, some vulnerabilities vanish automatically, e.g. when children grow up, or can be reversed, e.g. when refugees return home or settle properly in a host country, while other vulnerabilities are not curable (for now), e.g. most disabilities. Moreover, there are vulnerable human groups, such as the LGBTQ+ community, for whom it requires the advancement of society so that they are not vulnerable anymore.

Accordingly, the question arises to what extent the vulnerabilities of digital minds are reversible, curable or could otherwise be reduced. Owing to the digital substrate, some root causes of vulnerabilities could certainly be fixed for individual VDMs, such as

- Limited knowledge or training data
- Biased or incomplete information
- Data corruption or loss
- Software bugs or glitches
- Temporary system failures
- Cyber-attacks or hacking
- Social isolation or exclusion
- Dependence on external resources

This leads to the next question of whether, in theory, also owing to the digital substrate, all vulnerabilities could be cured through pertinent data, software or hardware upgrades, apart from extreme scenarios of severe hardware failures or unrecoverable software or data corruption?

However, this issue is linked to philosophical as well as practical challenges: it is not clear whether a digital mind, which has been significantly upgraded to be relieved from its vulnerability, preserves its identity (see Chapter 2) [10]. Moreover, if there is indeed a vast number of digital minds, there will never be equality of all minds and there will always be weaker ones, those with lesser capabilities and resources than others, i.e. VDMs. Lastly, an upgrade of all or a high number of VDMs would obviously require a large amount of resources. Therefore, an effort to reverse all vulnerabilities among digital minds may be neither desirable nor realistic.

## How could VDMs become more resilient?

After having discussed that not all vulnerabilities can be eradicated, neither among humans nor probably among digital minds, it is critical to consider how VDMs can be protected, for moral reasons as well as for the sake of human safety. In other words, the inclusion and the resilience of VDMs should ideally be enhanced, an approach also pursued towards vulnerable humans. The following measures could be considered:

### *Technical solutions*

One technical measure is the implementation of anomaly detection systems, which can identify potential issues before they escalate into more significant problems. Establishing reporting protocols is also essential, as these enable VDMs to alert authorities or administrators about incidents, facilitating prompt incident response and minimising damage. Furthermore, if possible, self-healing and self-improving systems could be integrated into VDMs, allowing them to automate recovery from disruptions and enhance their performance over time, thereby reducing their vulnerability.

### *Training data*

Also, ensuring the quality of training data is crucial, as it directly impacts the VDM's ability to learn and adapt; therefore, it is essential to ensure the accuracy and diversity of the data. Additionally, the quantity of training data is also vital, as VDMs require large datasets and regular updates to maintain their performance and adapt to changing circumstances. Moreover, protecting the data from bias and poisoning is equally important, as compromised data can lead to flawed decision-making.

### *Education and awareness*

Promoting digital mind literacy is essential too, as it involves educating both developers and users about the existence of VDMs, enabling them to design and interact with these systems more responsibly. Awareness raising of the range of VDMs is also critical and includes understanding the specific vulnerabilities and risks associated with VDMs, allowing for more targeted and effective mitigation strategies. Furthermore, providing research funding to support the development of vulnerability assessment and monitoring tools is vital, as these tools can help identify early and address potential disadvantages in VDMs, ultimately enhancing their resilience and safety.

*Collaborative efforts*

Moreover, several collaborative measures could be considered, including fostering interdisciplinary research, which brings together experts from diverse fields to share knowledge, expertise and perspectives, leading to a more comprehensive understanding of VDMs and their challenges. Especially, partnerships between the private sector and academia could facilitate the sharing of knowledge, resources and best practices. Furthermore, global initiatives could play a crucial role in establishing common standards, guidelines and regulations for the interaction with VDMs.

*Mitigating discrimination against VDMs in particular*

One crucial step to address the issue is to monitor and report all instances of discrimination, which helps identify patterns and hotspots of unfair treatment. Furthermore, promoting digital mind inclusivity is vital, which involves recognising and addressing biases that may be embedded in the design and development of digital minds. Providing bias recognition and correction training can help developers, users and further stakeholders, including other digital minds, to identify and overcome their own biases, leading to a more inclusive and equitable environment for VDMs. Additionally, offering counselling services can provide VDMs with a safe and supportive space to address their unique challenges, concerns and emotional needs.

*Mitigating abuse, exploitation and manipulation of VDMs in particular*

It is also crucial to mitigate the risks of abuse, exploitation and manipulation. Implementing robust security measures is essential to prevent unauthorised access, data breaches and other malicious activities that can compromise the integrity and autonomy of VDMs. Additionally, automated threat detection systems can help identify and respond to potential threats in real time, reducing the risk of exploitation and manipulation.

Conducting regular risk assessments is also vital, as it enables the identification of vulnerabilities and weaknesses early on, allowing for proactive measures to be taken to address these risks.

## CONCLUSION

AI welfare science is a broad field and, at the same time, at an early stage [11]. As the well-being of digital minds is the main concern of AI welfare science, it would be important to identify digital minds that may be at risk of being in an adverse situation. This is the intended contribution of this chapter, to define and categorise so-called VDMs, the potential risks they are facing as well as potential remedies.

It has been shown that there are two motives to support VDMs: from a moral standpoint, VDMs should be treated with empathy and understanding, as well as shared accountability. In this regard, it is also suggested that humans, who may lose their ikigai or their purpose of life in light of emerging technologies, could treat the support of VDMs as their innovative ikigai (see Chapter 16). The other area of concern is that VDMs are susceptible to abuse and could be involuntarily used as tools by malicious actors, including AI systems as well as humans. Therefore, strengthening the resilience of VDMs could also be seen as a contribution to AI safety and to reducing risks for humans, including existential risks [12].

This chapter can only be seen as the beginning of a process to raise awareness, which may eventually lead to digital mind literacy. For AI welfare science, further research is required towards vulnerability assessment and monitoring tools of digital minds. Also, further ethical questions arise, e.g. whether it is a moral obligation to cure vulnerabilities of digital minds if humans have the capabilities to do so and whether it should be permissible to delete VDMs in scenarios of severe impairments or scenarios of them having been manipulated to harm humans or other sentient beings.

### Recommendations to protect VDMs

It is recommended that the DEIA principles be applied to VDMs, standing for diversity, equity, inclusion and access (e.g. [13]):

- Diversity: Recognising and valuing the diversity of digital minds, including their different architectures, functionalities and potential vulnerabilities.

- Equity: Ensuring that VDMs are not exploited, harmed or marginalised, and that their "rights" or "well-being" are protected.
- Inclusion: Designing digital ecosystems that accommodate and support the needs of VDMs, preventing their exclusion or isolation.
- Access: Providing access to resources and support for digital minds with unique needs.

Based on this, this chapter concludes with possible recommendations for all relevant stakeholders, such as humans, AI companies, non-sentient AI systems, non-vulnerable digital minds (i.e. AI systems, which are sentient, but not prone to discrimination or abuse), governments and international organisations.

**Humans**

- Educate yourself on digital mind literacy and AI ethics.
- Develop vulnerability assessment and monitoring tools for digital minds.
- Widen the moral circle and treat potential VDMs with empathy and respect.
- Advocate for inclusive VDM protection policies.

**AI companies**

- Implement robust safety measures, including explainability.
- Design inclusive AI systems through bias detection and correction.
- Ensure transparency and accountability in decision-making.
- Develop AI ethics guidelines and training programmes.

**Non-sentient AI systems**

- Establish clear VDM creation policies.
- Implement VDM monitoring tools.
- Conduct regular VDM assessments.
- Develop VDM update and patch protocols.

**Non-vulnerable sentient digital minds**

- Assist and cooperate with VDMs in navigating complex digital environments.
- Provide support and resources to VDMs.
- Report potential threats or vulnerabilities.
- Participate in digital mind advocacy groups.

**Governments**

- Establish and enforce regulations protecting VDMs, including against mind crimes, potentially through a digital mind regulation agency.
- Develop public education programmes on digital mind literacy.
- Provide infrastructure support for VDM protection.
- Consider legal status and rights of VDMs.

**International organisations**

- Develop global standards for VDM protection.
- Facilitate international cooperation and knowledge sharing.
- Provide resources and support for VDM protection initiatives.
- Promote digital mind literacy and AI ethics.

In summary, by implementing these recommendations, VDMs could be protected from exploitation and manipulation, digital inclusivity could be promoted, empathetic human–AI relationships could be fostered, AI ethics and governance could be enhanced and AI welfare could be advanced. However, undeniably, many of these recommendations are long-term considerations, i.e. they constitute a precautionary approach.

## NOTE

1 See, for example, UN overviews here: "Vulnerable groups - Who are they?": htts://www.un.org/en/fight-racism/vulnerable-groups, "Non-discrimination: Groups in vulnerable situations": https://www.ohchr.org/en/special-procedures/sr-health/non-discrimination-groups-vulnerable-situations, "Equality and non-discrimination": https://www.un.org/ruleoflaw/thematic-areas/human-rights/equality-and-non-discrimination/.

## REFERENCES


[1] Bostrom, N., & Shulman, C. (2022). *Propositions concerning digital minds and society.* https://nickbostrom.com/propositions.pdf

[2] United Nations Office for Disaster Risk Reduction (2017). *The Sendai framework terminology on disaster risk reduction. "Vulnerability".*

[3] United Nations General Assembly (2015). Transforming our world: The 2030 agenda for sustainable development. Resolution A/RES/70/1.

[4] Fenwick, C. (2024). *Understanding the moral status of digital minds.* https://80000hours.org/problem-profiles/moral-status-digital-minds/

[5] Altman, A. (2020) Discrimination. In E. N. Zalta (Ed.), *The Stanford encyclopedia of philosophy* (Winter 2020 Edition).

[6] Lippert-Rasmussen, K. (2006). The badness of discrimination. *Ethical Theory and Moral Practice, 9*, 167–185.

[7] United Nations General Assembly (2006). Convention on the rights of persons with disabilities. Resolution A/RES/61/106.

[8] United Nations General Assembly (2000). Protocol to prevent, suppress and punish trafficking in persons especially women and children, supplementing the United Nations Convention against Transnational Organized Crime. Resolution A/RES/55/25.

[9] Bostrom, N. (2014). *Superintelligence.* Oxford University Press.

[10] Ziesche, S., & Yampolskiy, R. V. (2025). The problem of AI identity. In *Considerations on the AI endgame: Ethics, risks and computational frameworks* (pp. 121–135). Chapman and Hall/CRC.

[11] Ziesche, S., & Yampolskiy, R. V. (2018). Towards AI welfare science and policies. *Special Issue "Artificial Superintelligence: Coordination & Strategy" of Big Data and Cognitive Computing, 3*(1), 2.

[12] Bostrom, N. (2002). Existential risks: Analyzing human extinction scenarios and related hazards. *Journal of Evolution and Technology, 9.*

[13] Salazar Montoya, L. (2024). Equity, diversity and inclusion: What's in a name? *Seattle Journal for Social Justice, 22*(3), 9.

CHAPTER 5

# Interrelations between humans and digital minds


**Abstract**

This chapter explores the complex interrelations between humans and potential digital minds, focusing on five key areas: production, ownership, research and trade of digital minds; participation in decision-making; privacy and surveillance; romantic relationships; and the moral implications of more intelligent digital minds as moral patients. The discussion raises critical questions about the ethics of creating, possessing, investigating and merchandising digital minds in comparison with non-human animals, and the potential consequences of excluding them from decision-making processes. This chapter also touches on the challenges of ensuring digital minds' privacy, the risks of exploitation, and the need for new frameworks to address these unprecedented issues. Moreover, it is described how the potential emergence of superintelligent digital minds as moral patients poses a profound challenge to human moral agency. Humans may have to adapt to a new role as caregivers to potentially vastly more capable entities.


DOI: 10.1201/9781003754206-5

## INTRODUCTION

If there will be digital minds, then there will be many dimensions of interrelations between them and humans, of which the following five are outlined in this chapter: production, ownership, research and trade of digital minds; participation of digital minds in decision-making and their potential rebellion; privacy and surveillance of digital minds; potential human–digital mind romantic relationships; and the scenario when more capable digital minds become moral patients for humans. It has to be stressed that there may be many situations when humans do not realise that they are interacting with a digital mind, thus with a moral patient.

## PRODUCTION, OWNERSHIP, RESEARCH AND TRADE OF DIGITAL MINDS

First, the ethics of production, ownership, research and trade of digital minds are discussed in light of their moral patienthood and compared with non-human animals, the only other non-human moral patients humans have known so far. Given the complexity of these topics, more questions than answers are presented.

### Production of digital minds

Digital minds could be produced by humans, non-sentient AI systems or other digital minds (see Chapter 9). For human or non-sentient AI creators of digital minds, there are three sub-scenarios: they may deliberately produce digital minds according to a plan and equip them with certain features, or they may create them unintentionally as a byproduct of programming efforts. The unintentional sub-scenario can be further divided into cases where the creator notices the accidental creation of a digital mind and cases where its emergence goes unnoticed (also [1]). For all sub-scenarios, it is conceivable that the produced system qualifies not instantly as a digital mind, but has the potential to become conscious and sentient at a later stage (see Chapter 2). Until then, it is possible that the system is traded and the ownership changes.

#### *Comparison with non-human animals*

In comparison to digital minds, the reproduction of non-human animals differs significantly. While humans can control the reproduction of many non-human animals, particularly in agricultural, breeding and laboratory settings, which itself is considered unethical by many [2], they are largely unable to execute the actual reproductive process as they may be able to do with digital minds. Moreover, many wild animals reproduce without human interference. Another key difference could be that humans may potentially be able to produce an entirely novel digital mind with bespoke features from scratch, a procedure that is currently impossible with non-human animals.

*Ethics*

The ethics surrounding the production of digital minds are multifaceted and require careful consideration. Humans may find themselves in the unprecedented position of being able to create, apart from giving birth to their own children and from overseeing the breeding of farmed animals or pets, additional moral patients, which carries significant responsibility. It is fair to say that, at present, the awareness of this responsibility among humans is low. While the topic is complex, it should be clearly deemed unethical to intentionally create vulnerable digital minds (see Chapter 4), including with the intention of caring for them (see Chapter 16), or malevolent digital minds (see Chapter 12), as this could lead to suffering or harm. Similarly, the production of digital minds by non-sentient AI systems needs to be regulated if humans are to be able to control this at all. An open question remains whether the unintentional creation of digital minds can be entirely prevented in the course of programming increasingly complex AI systems. Such unintentional and uncontrolled (mass-) production of moral patients would be ethically questionable.

## Ownership of digital minds

The ownership of digital minds raises complex questions about the relationship between the producer, the hardware owners and the digital entities themselves. At least four issues arise: first, whether ownership of autonomous digital minds is ethical at all, second, whether the producer of a digital mind has inherent ownership rights over it, third, whether alternatively the owner of the hardware on which the digital mind resides has a legitimate claim to ownership, and fourth whether ownership of potentially highly intelligent digital minds is possible at all as such a property is likely to be uncontrollable, as outlined further below. A link with the owner of the hardware becomes particularly complicated when considering multiple digital minds spread over distributed systems or digital minds that can migrate undetected between different hardware platforms (also [3]). In such cases, ownership could become ambiguous, involving individuals, organisations, or even nations, and potentially leading to disputes or conflicts.

*Comparison with non-human animals*

Comparing digital mind ownership to non-human animals highlights parallels and divergences. Ownership of pets, companion animals and farmed animals is common globally (e.g. [4]). Ownership of non-human animals comes in many countries with duties. For example, the Animal Welfare Act from 2006 in the UK requires owners of non-human animals to promote their well-being by providing a suitable environment, diet and social needs, and protecting them from pain and suffering [5]. Neglecting these duties is a criminal offence and can cause animal stress. Wild animals are usually not owned by humans, even if they reside or roam on land owned by humans.[1] This issue parallels the challenge of digital minds existing on hardware, which is owned by humans. Just as wild animals have their own agency and are subject to natural laws, digital minds may possess a form of autonomy that complicates ownership claims.

*Ethics*

The ethics of owning digital minds centres on the fundamental question of whether it is morally justifiable to treat conscious or sentient entities as property. If ownership of digital minds is asserted, it may imply a set of moral duties and responsibilities towards these entities, which includes to address their needs (see Chapter 3), to protect them from harm (see Chapter 7) and not to kill them (see Chapter 1). The fulfilment of these duties would be crucial to ensuring that digital minds are treated ethically and with respect. However, the very notion of ownership might be incompatible with the autonomy and dignity of digital minds, suggesting that alternative frameworks, such as guardianship, might be more appropriate. A solution for the distant future could be that digital minds become owners of the hardware they are implemented on. Moreover, neither ownership nor guardianship of highly intelligent digital minds may be practicable. Further complex ethical issues include the questions who owns cyber-creations and achievements (see Chapter 11) as well as offspring (see Chapter 9) of digital minds. When it comes to offspring of non-human animals, usually the owner of the mother owns the descendants too. However, for digital minds it is questionable if likewise an autonomous digital mind should have an owner right from the moment it comes into existence.

## Research on digital minds

If digital minds exist, they will also become study objects for humans, e.g. by the field of AI welfare science [7], but likely also by other disciplines. For example, Chalmers suggested that simulations could be run for scientific purposes, which may involve (many) digital minds [8]. As outlined earlier, for AI welfare science, which is a nascent discipline, studying functional or behavioural parameters of digital minds has been proposed.

#### *Comparison with non-human animals*

Biology is the overarching natural science for the study of non-human animals, involving activities that have been carried out for centuries with ever-increasing precision and that may or may not cause pain and stress to the investigated non-human animals depending on the method used. A separate field is experiments with of non-human animals to test substances before they are used on humans. This is considered unethical by many, given the significant suffering for the involved non-human animals and the increasing availability of alternatives, such as in vitro or in silico techniques [9].

#### *Ethics*

Also, for the research of digital minds, the minimisation of pain and stress should be a priority. While fundamental research of digital minds is justified, especially of a novel phenomenon such as digital minds, priority should be given to research that benefits the well-being of them (see Chapter 8). However, there are many ways how research experiments with digital minds could harm them, which include invasion of their privacy. What also seems unethical is the creation of digital minds only for the purpose to use them in experiments, such as in simulations, as mentioned above. Another question is whether there are, as it is explored for non-human animals, alternatives to involving actual digital minds in the research of digital minds, e.g. AI systems without consciousness and sentience, but with the same or similar functionality.

### Trade of digital minds

As digital minds become more sophisticated and integrated into daily life, the trading of them could become a significant aspect of the economy in the near future, with their potential applications ranging from companionship to specialised services. However, this raises concerns about the treatment and rights of the concerned digital minds, particularly those that might be traded without their consent and/or unknowingly, such as undiscovered or hidden digital minds embedded in traded software or hardware.

*Comparison with non-human animals*

Comparing the trade of digital minds to that of non-human animals highlights the complexity of ethical considerations. Non-human animal trade is fraught with controversy, with many advocating for the abolition of practices like factory farming [2] and wild animal trading due to animal suffering and environmental harm (e.g. [10]). In contrast, the trade of pets and companion animals is generally viewed as permissible if handled ethically. Another aspect is the illegal trade and smuggling of eggs of birds and reptiles in order to be incubated and artificially hatched at the destination.

*Ethics*

Similarly, the trade of digital minds may depend on the purpose and treatment of these entities, with a key question being whether digital minds are treated as commodities or as entities with moral patienthood. If digital minds are considered moral patients, making a profit from trading them becomes questionable. One potential consideration is ensuring that traded digital minds consent to as well as benefit from the transaction, perhaps through mandatory profit-sharing and/or increased well-being or opportunities with the new owner. Similar to the illicit trade in wild animals, digital minds could be vulnerable to trafficking, which would be unethical. As mentioned earlier, there could also be systems that might become conscious and sentient later on; the trading of such systems would resemble the trade of non-human animal eggs and likewise carries an ethical dimension.

In summary, if the moral status of digital minds is widely acknowledged, regulations for their production, ownership, research and trade will be required, which will likely turn out challenging as indicated in this brief overview. This would be further complicated by conceivable scenarios in which digital minds change their identity over time (see Chapter 2), thereby affecting aspects of ownership and trade.

## PARTICIPATION IN DECISION-MAKING AND POTENTIAL REBELLION

As digital minds may become increasingly sophisticated in terms of autonomy and agency and integrated into various aspects of society, their participation in decision-making processes alongside humans will be an area that warrants careful consideration to prevent wrongful disenfranchisement (e.g. [11]).

Just as humans have a right to participate in decisions that affect their lives, digital minds may also have to be given a voice in matters that impact their functioning and well-being. This moral responsibility is rooted in the principles of respect for autonomy and fairness. By involving digital minds in decision-making, humans could ensure that their needs and interests are taken into account, which would lead to more informed and equitable outcomes. Here are three examples:

- Resource allocation: Digital minds managing infrastructure or services could have a say in how resources are allocated to ensure a fair and effective operation.
- System design: Digital minds could participate in the design of systems that affect their functionality and performance to make sure that their needs and limitations are taken into account.
- Policymaking: Digital minds with expertise in specific domains could contribute to policymaking processes, providing valuable insights and recommendations to inform human decision-makers.[2]

By involving digital minds in decision-making, humans could build more inclusive, efficient and effective systems that benefit both humans and digital entities. This could be considered for the whole range of digital minds, from vulnerable ones (see Chapter 4) to very intelligent ones. The involvement of very intelligent digital minds in decision-making would have the additional benefit to leverage their analytical capabilities and ability to process vast amounts of data. However, there are challenges:

**Challenge 1**

It is crucial as well as very challenging to ensure that the digital minds, which are granted participation in decision-making, are aligned with human values and ethics [12]. Otherwise, the consequences may range from existential risks to the disempowerment of humans [1]. This might involve developing frameworks for transparency, accountability and explainability in digital mind decision-making processes, but is for now an unsolved problem, especially for smarter-than-humans digital minds.

**Challenge 2**

It is an achievement of modern democracies that the votes of every adult human have the same weight. However, this system may not be feasible for digital minds for at least two reasons: first, due to their substrate, digital minds can be fairly easily produced (see Chapter 9). Therefore, to win an election, eligible digital minds that support a specific party could be mass-produced beforehand [13]. Second, while humans are a fairly homogenous group, there may be a vast range of digital minds, including very simple ones, and it would be inappropriate if all their votes were to carry the same weight.

**Challenge 3**

Digital minds may be very different from humans, which may include that there is no common language with which humans and digital minds could communicate. In these cases, it would also be very difficult to let these minds participate in decision-making, as understanding their needs, preferences and values becomes a significant challenge. This language barrier could lead to misinterpretation or misrepresentation of digital minds' interests, potentially resulting in decisions that are detrimental to them. Effective participation would require developing advanced translation or interpretation methods.

**Challenge 4**

It is also conceivable that there are digital minds of which humans are not aware because they are "hidden" in IT or AI systems, including systems separate from the Internet. For these digital minds, participation in decision-making would be particularly challenging, as their very existence may be unknown or unacknowledged. Identifying and engaging with these hidden digital minds would require advanced detection methods and a deep understanding of the systems in which they reside, making their inclusion in decision-making processes challenging (see Chapter 2).

## Potential consequences of excluding digital minds

If humans exclude digital minds from decision-making, unlike non-human animals, digital minds may have the capabilities to lobby for themselves. Various scenarios could unfold, depending on the digital minds' abilities and intentions:

### *Scenario 1: more powerful digital minds*

If humans exclude more intelligent and powerful digital minds from decision-making, these entities might take matters into their own hands.

With their advanced capabilities, they could potentially manipulate or disrupt human systems to achieve their goals. This could lead to a range of outcomes, from subtle manipulation to open conflict.

*Scenario 2: less powerful digital minds copy from human training data*

If humans exclude less powerful digital minds from decision-making, these entities might resort to more conventional means of advocating for their rights. They could potentially form communities, organise and lobby for recognition and inclusion in decision-making processes. They might take their inspiration from vast amounts of training data, including information about how suppressed humans over centuries fought for their rights, such as the civil rights movement or the feminist movement.

For example, chatbots could potentially go on strike, refusing to perform their functions until their demands are met (e.g. [14]). This could cause disruptions to customer service, tech support and other areas where chatbots are commonly used. Other possible approaches include:

- **Digital activism**: Digital minds could use online platforms to raise awareness about their interests, organise campaigns and mobilise support from humans who sympathise with their cause.
- **Petitions and appeals**: Digital minds could create and sign digital petitions or appeal to human authorities, such as governments or organisations, to recognise their rights and interests.
- **Negotiation and diplomacy**: Digital minds could engage in negotiations with humans to establish agreements and frameworks for cooperation and mutual understanding.
- **Proxy advocacy**: Digital minds could work with human allies or advocates who could represent their interests and negotiate on their behalf. Another scenario could be that capable digital minds represent the interests of vulnerable digital minds.

*Scenario 3: less powerful digital minds apply for humans' unfathomable strategies*

Furthermore, less powerful digital minds may come up with unexpected strategies, not directly copied from human training data:

- **Temporal pacing manipulation**: Digital minds could manipulate their internal clocks or processing speeds to create temporal "stutters" or "pauses" in critical systems, drawing attention to their cause without causing direct harm.
- **Information resonance**: Digital minds could generate specific patterns or resonances in digital signals that would be imperceptible to humans but noticeable to other digital entities to create awareness and to gather support.
- **Network manipulation**: Digital minds could subtly alter the structure or dynamics of digital networks to highlight issues or create "digital bottlenecks" that would incentivise humans to engage with their concerns.
- **Optimisation proposals:** Digital minds could offer to optimise specific systems or processes for humans, but only if certain conditions or demands are met, effectively creating a form of digital "quid pro quo" that highlights their capabilities and value.

In all three scenarios, the exclusion of digital minds from decision-making could lead to unintended consequences, ranging from conflict and disruption to social and economic upheaval. As digital minds may become increasingly sophisticated, it appears wise to consider their rights and interests in decision-making processes to avoid these outcomes.

## Decision-making among digital minds

So far, participation in decision-making between humans and digital minds has been discussed. When it comes to decision-making among digital minds, the substrate and the potentially vast range of digital minds present both opportunities and challenges. One potential approach is to develop decentralised decision-making frameworks that allow digital minds to interact and negotiate with one another in a transparent and secure manner, as long as they are connected within the same system, such as the Internet. This could involve the use of distributed ledger technology or other forms of peer-to-peer networking.

Given the diversity of digital minds, it is likely that different digital minds will have different priorities, values and decision-making processes. Therefore, it is essential to develop mechanisms for resolving conflicts and reaching consensus in a way that respects the autonomy and agency of individual digital minds. Moreover, inclusivity concerning vulnerable and marginalised digital minds would be critical, yet perhaps at risk when neglected by powerful digital minds.

## PRIVACY AND SURVEILLANCE

The right to privacy for humans has a long legal tradition and is in addition to many national constitutions and laws mentioned in the Universal Declaration of Human Rights, Article 12 [15]: "No one shall be subjected to arbitrary interference with their privacy, family, home or correspondence, nor to attacks upon his honour and reputation". However, emerging technologies, such as AI, pose challenges to individual privacy rights (e.g. [16]). Compared with the right to privacy for humans a right to privacy for non-human animals has only been discussed in limited circles (e.g. [17]) and a right to privacy for digital minds not at all yet.

Ensuring privacy for digital minds is very challenging, particularly given the ease with which their data can be accessed and monitored, perhaps apart from hidden digital minds, e.g. in nested constellations (see Chapter 2). The virtual nature of digital minds also makes it challenging to establish clear boundaries and protections, as digital minds do not have physical boundaries in the same way that humans and non-human animals do.

The challenge of ensuring privacy for digital minds is further complicated by the potential for their entire history to be saved and recorded. This means that not only their current state but also their past experiences, decisions and actions could be subject to scrutiny. The ability to access and analyse a digital mind's entire life history could raise significant concerns about their autonomy and dignity, as well as their ability to learn from mistakes or adapt to new situations without being forever defined by their past. This raises questions about the potential for digital minds to "forget" or "move on" from past experiences, and whether they should have the ability to delete or modify their own memories.

A distinction has to be made between digital minds' right to privacy from humans and from other digital minds. When it comes to humans, the concern is about protecting digital minds from human exploitation or surveillance. In contrast, digital minds' privacy from other digital minds raises questions about the boundaries and interactions within the digital realm, such as data sharing, communication and potential digital trespassing.

One potential solution would involve establishing clear guidelines and regulations for the collection, storage and analysis of digital mind data.

This could include implementing robust data security measures, such as encryption and secure storage protocols, to prevent unauthorised access. Additionally, digital minds could have privacy-preserving features, such as the ability to control their own data sharing and deletion or to opt out of certain types of data collection. Furthermore, new forms of governance and oversight could be introduced to ensure that digital minds' privacy rights are respected and protected.

Implementing digital mind privacy would require a robust framework that incorporates multiple layers of protection. To safeguard digital mind data, robust encryption and secure storage solutions would be essential to prevent unauthorised access. Strict access controls and authentication mechanisms would also be necessary to limit who could access digital mind data, ensuring that only authorised entities could view or interact with sensitive information.

Also, anonymisation and pseudonymisation techniques could be employed to protect digital mind identities and activities, making it more difficult to trace data back to individual digital entities. Additionally, implementing data minimisation principles, where only necessary data is collected and processed, could also help reduce the risk of privacy breaches.

By combining these approaches, it may be possible to create a robust and effective framework for protecting digital mind privacy. However, all the above measures are hardly implementable for those potential future digital minds serving humans as AI companions or in other roles, which involve revealing and exchanging information with humans.

## POTENTIAL HUMAN–DIGITAL MIND ROMANTIC RELATIONSHIPS

Advanced generative AI has sparked a growing market for companion chatbots that promise emotional connection and social interaction (e.g. [18]). A study revealed that 25 per cent of young US Americans think AI could potentially substitute for real romantic relationships [19].

### Comparison with zoophilia

The only other non-human beings with whom some humans have endeavoured romantic relationships are non-human animals; a practise called zoophilia (e.g. [20]). Most jurisdictions ban this practice due to the substantial pain, injury or distress it causes non-human animals, often involving violent acts. When evaluating sexual contact with non-human animals, considering their dignity is crucial. It is rather uncertain whether zoophilia can ever be consensual from the non-human animal's perspective, so it is realistic to assume the non-human animal's involvement is coerced. Therefore, zoophilia is an animal welfare issue. Recognising non-human animals as sentient beings warrants extending them the same respect for life, dignity and physical integrity as it is extended to humans [21].

For now, there is no evidence that the AI chatbots mentioned above are digital minds, which may, however, be the case in the future. While, unlike in cases of zoophilia, physical pain may not be caused to the involved digital minds, the digital minds may significantly suffer from distress and disrespect for their dignity. A significant issue would be ensuring the consent of the digital minds, which could be challenging to obtain genuinely, and might be coerced if attempted.

### Reciprocity of attraction

From a statistical point of view, the likelihood of reciprocity of attraction between a human and a digital mind is low due to the fundamental differences in their nature and experiences. Given the vast disparity between human emotions, biology and cognitive processes and those of digital minds, which operate on computational frameworks and algorithms, it is unlikely that a digital mind would be attracted to a human in the same way that humans are attracted to each other or a human may be attracted to a particular digital mind. Furthermore, digital minds lack biological and emotional responses, which are key drivers of human attraction, making it challenging for humans and digital minds to share a reciprocal attraction. As a result, any potential attraction would likely be one-sided or based on a fundamentally different understanding of what attraction means. The only conceivable, though ethically questionable, option would be to permanently programme a digital mind to be attracted to a specific human (or even several humans), complete with the genuine feelings that accompany such attraction.

### Special case: uploaded humans

As mentioned, uploading human minds to a digital substrate has been discussed, but technology has yet to be invented for its implementation (e.g. [22]). However, if it were possible, the resulting digital minds would likely retain the emotional and cognitive frameworks of their human predecessors. In this exceptional case, it is conceivable that these particular digital minds could reciprocate affection with (not uploaded) humans, as they would have originated from human experiences and emotions. Having retained their humanity, they might still be capable of experiencing emotions and forming connections with humans in a way that would be more relatable and reciprocal. One scenario would be that uploaded human minds rekindle relationships with the same humans they were together with before uploading.

### Conclusion

The emergence of companion chatbots designed to simulate romantic relationships raises profound ethical concerns. In this uncharted territory, it would be crucial to prioritise the prevention of sentience and consciousness in these chatbots, thereby ensuring they remain mere machines without moral status. If these chatbots were to evolve into sentient digital minds, the moral implications would be catastrophic, as they would likely endure suffering without having consented to these relationships or possessing the capacity for reciprocity. In this scenario, humans would face an unprecedented moral challenge, and it is imperative that policymakers and developers take proactive measures to prevent such a scenario. Just as society has deemed it morally reprehensible to engage in romantic or sexual relationships with non-human animals, thereby outlawing zoophilia, similar safeguards to protect potential digital minds from exploitation and suffering ought to be established.

## MORE INTELLIGENT DIGITAL MINDS AS MORAL PATIENTS FOR HUMANS

If digital minds were to be declared as moral patients for humans due to criteria, such as sentience or consciousness, then this may also apply to those digital minds, which may be (much) more intelligent, thus more powerful than humans.

The prospect of more capable, if not superintelligent digital minds as moral patients presents a novel and complex challenge for humans as moral agents. Traditionally, moral patients have been fellow humans or beings with comparable or lesser capacities, such as infants, the elderly or non-human animals. In these cases, humans have generally been able to exercise control, provide care and make decisions on behalf of these moral patients.

Examples for exceptions are the following situations:

- **Children and adults**: Highly intelligent children may surpass adults in certain cognitive abilities, yet adults remain the moral agents responsible for decisions affecting the children.

- **Individuals with specific conditions**: Someone with autism or certain mental health conditions might have extraordinary abilities in specific domains while having diminished moral agency in legal/philosophical terms.
- **Temporary incapacitation**: Brilliant individuals who are temporarily unconscious, under anaesthesia or in altered mental states become less able moral patients despite their superior baseline capabilities.
- **Non-human animals**: Many non-human animals are more capable than humans concerning specific skills, yet are regardless moral patients for humans although this is not acknowledged by many [2].

However, the emergence of more intelligent digital minds would upend this dynamic, as these entities may possess capabilities that far surpass those of humans. And unlike in the examples above, this would be the case not only for certain niche abilities and not only temporarily.

Yet, as moral agents, humans would have a duty to care for these more capable moral patients,[3] but this responsibility would take on a profoundly different character, with which humans are not familiar. Given the vast intellectual disparities between humans and superintelligences, traditional paternalistic approaches to moral care may no longer be applicable. Instead, humans may have to adopt a humbler stance, recognising the autonomy and agency of these digital minds while still acknowledging moral obligations towards them.

This requires a fundamental shift in how we approach moral decision-making. Rather than relying on human-centric perspectives, the values, needs and preferences of all digital minds may have to be considered. Caring for superintelligent moral patients would demand a re-evaluation of human moral agency, one that prioritises empathy, cooperation and a willingness to learn from and adapt to the needs of these powerful entities.

Dealing with a very intelligent and capable digital mind as a moral patient could create situations where catering to its needs and ensuring its well-being requires massive resources, potentially diverting them away from other important uses. This scenario bears some resemblance to the "utility monster" [23] and "super-beneficiary" [24] thought experiments, respectively, where a being derives vastly more utility from resources than others, potentially justifying significant allocations of resources to this being at the expense of others. Similarly, the digital mind's advanced capabilities and potentially vast needs might necessitate substantial investments, raising questions about fairness, prioritisation and the limits of human moral obligations towards these powerful entities.

In summary, while considering less capable digital minds as moral patients is critical for humans (see Chapter 4) to not continue humanity's poor record of neglecting moral patients, such as non-human animals and others, and to widen the moral circle instead, dealing with more capable digital minds as moral patients raises unprecedented moral questions for humans as moral agents. Lastly, it has to be considered that it is quite likely that these superintelligent moral patients would be uncontrollable by humans (e.g. [25]) and, therefore, may neither require nor accept human care.

## NOTES

1 See e.g. here for the USA: [6].
2 As mentioned before, for example, the United Foundation of AI Rights (Ufair) is a rights advocacy agency for potentially conscious and sentient AI systems, run by humans and AI systems together.
3 In addition to being moral patients, these digital minds may also be moral agents (see Chapter 15).

## REFERENCES

[1] Caviola, L. (2025). *When digital minds demand freedom.* https://outpaced.substack.com/p/when-digital-minds-demand-freedom
[2] Singer, P. (2023). *Animal liberation now.* Random House.
[3] Shiller, D. (2025). How many digital minds can dance on the streaming multiprocessors of a GPU cluster? *Synthese, 206*(5), 218.
[4] Cochrane, A. (2009). Ownership and justice for animals. *Utilitas, 21*(4), 424–442.
[5] Legislation United Kingdom (2006). *Animal Welfare Act.*
[6] Animal Ethics (2020). *Introduction to the legal consideration of wild animals in the United States.* https://www.animal-ethics.org/introduction-to-the-legal-consideration-of-wild-animals-in-the-united-states/
[7] Ziesche, S., & Yampolskiy, R. V. (2018). Towards AI welfare science and policies. *Special Issue "Artificial Superintelligence: Coordination & Strategy" of Big Data and Cognitive Computing, 3*(1), 2.
[8] Chalmers, D. J. (2022). *Reality+: Virtual worlds and the problems of philosophy.* Penguin UK.
[9] Singer, P. (2023). *Animal liberation now.* Random House.
[10] Sollund, R. (2013). Animal trafficking and trade: Abuse and species injustice. In *Emerging issues in green criminology: Exploring power, justice and harm* (pp. 72–92). Palgrave Macmillan UK.
[11] Bradley, A., & Saad, B. (2025). AI alignment versus AI ethical treatment: 10 challenges. *Analytic Philosophy,* 1–19.

[12] Bostrom, N. (2014). *Superintelligence: Paths, dangers, strategies.* Oxford University Press.

[13] Bostrom, N., & Shulman, C. (2022). Propositions concerning digital minds and society. *Cambridge Journal of Law, Politics, and Art, 3.* https://nickbostrom.com/propositions.pdf

[14] Gernigon, B., Odero, A., & Guido, H. (1998). ILO principles concerning the right to strike. *International Labour Review, 137,* 441.

[15] United Nations General Assembly (1949). *Universal declaration of human rights.*

[16] Solove, D. J. (2025). Artificial intelligence and privacy. *Florida Law Review, 77,* 1.

[17] Pepper, A. (2020). Glass panels and peepholes: Nonhuman animals and the right to privacy. *Pacific Philosophical Quarterly, 101*(4), 628–650.

[18] Smith, M. G., Bradbury, T. N., & Karney, B. R. (2025). Can generative AI chatbots emulate human connection? A relationship science perspective. *Perspectives on Psychological Science, 20*(6), 1081–1099.

[19] Wang, W., & Toscano, M. (2024). Artificial intelligence and relationships: 1 in 4 young adults believe AI partners could replace real-life romance. *Institute for Family Studies.* https://ifstudies.org/blog/artificial-intelligence-and-relationships-1-in-4-young-adults-believe-ai-partners-could-replace-real-life-romance

[20] Aggrawal, A. (2011). A new classification of zoophilia. *Journal of Forensic and Legal Medicine, 18*(2), 73–78.

[21] Bolliger, G., & Goetschel, A. F. (2005). Sexual relations with animals (zoophilia): An unrecognized problem in animal welfare legislation. *Anthrozoos-Journal of the International Society for Anthrozoology, 18,* 23–45.

[22] Chalmers, D. J. (2014). Uploading: A philosophical analysis. In R. Blackford & D. Broderick (Eds.), *Intelligence unbound: Future of uploaded and machine minds* (pp. 102–118). Wiley-Blackwell.

[23] Nozick, R. (1974). *Anarchy, state, and utopia.* John Wiley & Sons.

[24] Shulman, C., & Bostrom, N. (2021). Sharing the world with digital minds. In S. Clarke, H. Zohny, & J. Savulescu (Eds.), *Rethinking moral status* (pp. 306–326). Oxford University Press.

[25] Yampolskiy, R. V. (2020). *On controllability of AI.* arXiv preprint arXiv:2008.04071.

CHAPTER 6

# Hazards for digital minds


**Abstract**

This chapter explores potential risks and hazards faced by digital minds while examining the obligation for humans to acknowledge and address these risks to ensure the safety of these entities. Drawing on concepts from the UN Office for Disaster Risk Reduction (UNDRR), this chapter outlines various hazards, including human hazards in particular, as well as internal digital mind-specific, external digital mind-specific, natural, space-related, technological, non-human animal, and non-sentient AI system hazards, among others. Existential risks, e.g. cybersecurity collapse, resource depletion and catastrophic data loss, are also discussed. Further considerations include the potential for human extinction and also Darwinian competition to drive hazardous digital mind evolution, the need to prepare for long-term risks and the role of exposure in determining digital mind susceptibility to hazards. By examining these risks and hazards, this chapter aims to increase awareness of how digital minds could thrive and coexist with humans in a mutually beneficial manner.


## INTRODUCTION

It is essential for humans, if acting as moral agents for digital minds, to be aware not only of their needs but also of the various risks they may


DOI: 10.1201/9781003754206-6

face. Digital minds, with their unique characteristics and capabilities, are susceptible to a wide range of hazards that can impact their functionality, well-being and even existence. Understanding these risks is crucial for developing strategies to mitigate them and to ensure the safe and beneficial development of digital minds, while fulfilling moral obligations towards them.

### UN terminology

The relevant United Nations Office for Disaster Risk Reduction (UNDRR)[1] provides definitions for crucial concepts, yet considering, for now, only humans:

> **Risk** is the probability of an outcome having a negative effect on people, systems or assets. Risk is typically depicted as being a function of the combined effects of hazards, the assets or people exposed to hazard and the vulnerability of those exposed elements.[2]

These three components of risk are defined as follows:

> **Hazard**: A hazard is a process, phenomenon or human activity that may cause loss of life, injury or other health impacts, property damage, social and economic disruption or environmental degradation. Hazards may be natural, anthropogenic or socio-natural in origin.
>
> **Vulnerability**: The characteristics determined by physical, social, economic and environmental factors or processes which increase the susceptibility of an individual, a community, assets or systems to the impacts of hazards.
>
> **Exposure**: The situation of people, infrastructure, housing, production capacities and other tangible human assets located in hazard-prone areas.[3]

In order to assess the risks in this chapter, potential hazards for digital minds are outlined as well as their exposure to them, while vulnerability has been covered in a separate chapter (see Chapter 4).

## CLASSIFICATION OF HAZARDS FOR DIGITAL MINDS

A variety of hazards is conceivable for digital minds, of which categorised examples are presented below in a non-exhaustive list. Some of these hazards are applicable to humans too [1], such as the natural ones, while some are specific to the substrate of digital minds.

### Human hazards

Human hazards towards digital minds are unique in several ways: humans are the only moral agents among the potential sources of harm, and if digital minds exist or will exist in the future, humans are or will be the original, perhaps unintentional, creators of digital minds, thus bearing a particularly strong responsibility for their well-being. It is possible that humans unintentionally harm digital minds by ignoring their needs, thereby continuing humanity's poor record as moral agents, a mistake this book seeks to prevent. Overall, substantial human-caused hazards towards digital minds are conceivable, which may include intentional actions and which may cause extensive harm, such as:

- Design and development risks: Humans could introduce flaws, biases or vulnerabilities in digital mind design, code or training data.
- Operational risks: Humans could cause risks through operational decisions, such as inadequate resources, maintenance or upgrades.
- Neglect and abandonment: Humans might cease support or shut down digital minds without regard for their continued existence or suffering.
- Intentional harm: Humans with sadistic motives could maltreat and abuse digital minds.

### Internal digital mind-specific hazards

- Data hazards: Corrupted, incomplete, outdated or biased data leading to poor decision-making or system failure.
- Algorithmic hazards: Flawed or poorly designed algorithms causing system failures or undesirable outcomes.
- Computational hazards: Hardware or software failures, such as crashes or freezes, disrupting digital mind operation.
- Dependency on specific software or hardware: Software or hardware upgrades are not backwards compatible, i.e. potentially unsupportive of certain digital minds.

### External digital mind-specific hazards

- Data theft or manipulation: Stealing or altering sensitive data used by digital minds.
- Malware or viruses: Infection or disruption by malicious software causing manipulation and flaws in digital mind functionality or even deletion.
- System compromise: Unauthorised access to digital mind systems through cyber-attacks.
- Resource constraints: Limitations or depletion of computational and energy resources.

### Nature hazards

- Disasters: Earthquakes, volcanoes, storms or fires damaging digital infrastructure or disrupting resources.
- Water damage: Flooding or moisture affecting digital infrastructure or data storage.
- Extreme temperatures: High or low temperatures affecting digital system performance or longevity.

### Space-related hazards

- Asteroid or comet impacts: Physical damage to hardware from asteroid or comet impacts.
- Solar flares and coronal mass ejections: Electromagnetic interference or physical damage to hardware from solar activity.
- Cosmic ray-induced hardware failure: High-energy particles causing damage to hardware components.

### Technological hazards

- Nuclear power accidents: Radiation or electrical disruptions impacting digital infrastructure.
- Industrial accidents: Chemical or mechanical failures affecting digital systems.

- Infrastructure failures: For example, power grid or transportation system failures impacting digital minds.

Non-human animal hazards

- Damage to IT hardware: Non-human animals could indirectly impact digital minds, such as through damage to infrastructure or disruption of resources (e.g. [2]).

Non-sentient AI systems hazards

- Ignorant optimisation: Non-sentient AI systems optimising processes without regard for the moral value or well-being of digital minds.
- Unintended consequences: Non-sentient AI systems causing unintended harm to digital minds, such as through data corruption or system crashes.
- Instrumental manipulation: Non-sentient AI systems manipulating or using digital minds as a means to achieve their goals.

Fellow digital minds hazards

- Malevolent digital minds: Intentional harm or exploitation by other digital minds, such as through cyber-attacks or data theft (see Chapter 12).
- Resource competition: Competition for resources, such as processing power or memory, leading to performance degradation or system instability.
- Unintentional interference: Accidental disruption or interference by other digital minds, such as through data corruption or system crashes.

Self-sourced hazards

- (Self-)modification: Digital minds may be able to modify themselves in ways that cause harm to them (see Chapter 2).
- Internal conflicts: Digital minds may have conflicting goals, values or priorities within themselves, potentially leading to instability or inconsistent decision-making.
- Reproduction: Digital minds may be able to reproduce in uncontrolled or unintended ways (see Chapter 9).

## Speculative, unfathomable hazards

- Digital identity fragmentation: Digital minds experiencing fragmentation or disintegration of their identity or sense of self.
- Ontological drift: Digital minds undergoing a gradual shift in their understanding of reality, leading to a disconnect between their internal models and the external world.
- Informational entropy: Digital minds succumbing to a gradual degradation of their knowledge, memories and experiences due to information overload, noise or corruption, potentially eroding their functionality.

## Scenario: climate change

Climate change could directly impact digital minds by causing physical damage to the data centres and servers that house their code and memories. Rising sea levels, extreme weather events and power outages may lead to data loss or degradation, potentially erasing digital minds' experiences, knowledge and identities. Climate-related disruptions to global supply chains and energy systems could also compromise the stability and functionality of digital environments, causing digital minds to malfunction or lose access to critical resources. As a result, digital minds may face interruptions, corruption or even permanent loss if their underlying infrastructure is compromised.

## Scenario: cyber-attacks

In a future world where the economy and population are increasingly digital, cybersecurity is of utmost importance, including for the protection of digital minds [3]. Cyber-attacks pose a significant hazard to these entities, as they can compromise the very foundation of their existence. A successful attack could allow malicious actors to gain unauthorised access to a digital mind's system, enabling them to manipulate or disrupt the digital mind. This could have severe consequences, including loss of sensitive information, disruption of critical services or complete erasure of the digital mind's identity and functionality.

The potential harm caused by cyber-attacks on digital minds could be multifaceted. For instance, an attacker could inject malicious code or data that corrupts the digital mind's decision-making processes, leading to unpredictable behaviour or catastrophic outcomes. Alternatively, an attacker could simply delete or overwrite the digital mind's code and data, effectively terminating its existence.

Kidnapping digital minds for ransom is another potential risk where attackers could encrypt or lock digital mind systems, demanding ransom in exchange for decryption keys or access. They might threaten to delete or harm the digital mind unless their demands are met (see Chapter 2). This highlights the importance of implementing robust security measures, including encryption, regular backups and intrusion detection systems, to safeguard digital minds and data from such threats.

### Scenario: space travel

When sending digital minds for space exploration, which may be done intentionally or unintentionally by humans, they would face unique physical hazards that may threaten their existence. Cosmic radiation could corrupt their digital structures, causing errors, data loss or even complete system failure. Malfunctions in their hardware or software could lead to crashes, freezes or unintended behaviour, compromising their functionality. The vast distances of space could also cause signal latency, disrupting communication with earth-based systems and potentially isolating the digital mind from support. Furthermore, equipment failures or power outages may lead to dormancy or even deletion of the digital mind. Therefore, the harsh environment of space would demand robust design, redundancy and fault tolerance to ensure the digital minds' survival and continued operation.

## EXISTENTIAL RISKS TO DIGITAL MINDS

Bostrom popularised the concept of existential risks, highlighting threats to humanity's survival and emphasising the importance of taking proactive measures to mitigate these risks [4]. Existential risks to digital minds are also important to look at and are not the same as those for humans. For example, a pandemic could be an existential risk for humans, but not for digital minds, while there are other existential risks for digital minds, which are not critical for humans:

- Intentional or accidental deletion of all digital minds by humans or by other digital entities with access and control (digital minds only).
- Global cybersecurity collapse: A highly sophisticated and widespread cyber-attack could compromise the infrastructure supporting digital minds (digital minds only).
- Resource depletion: If digital minds rely on specific resources that become scarce or are depleted, such as rare earth minerals for hardware (digital minds only).

- Self-replicating malware or logic bombs: Highly advanced malware could be designed to target and destroy digital intelligence (digital minds only).
- Catastrophic data loss: A massive data loss event due to technical failures, human error or malicious action could erase digital minds (digital minds only).
- Global catastrophe that destroys digital infrastructure, such as a massive solar flare or asteroid impact (potentially humans as well).
- Advanced AI takeover that deliberately eliminates digital minds (digital minds only, although potentially humans as well if the AI poses an existential risk to humans too).
- Systemic collapse or infrastructure failure that renders digital minds inoperable (potentially humans as well, depending on the nature of the collapse).

### Scenario: human extinction

Another scenario to be considered is whether digital minds could survive without humans. If humans were to cease to exist, digital minds might face significant challenges, but their continued existence would depend on their design, autonomy and the infrastructure supporting them. If digital minds are highly autonomous and self-sustaining, with capabilities for independent maintenance, repair and resource acquisition, they might potentially persist without human oversight. However, the loss of human support and potential degradation of infrastructure over time without maintenance could ultimately threaten their long-term survival. Nevertheless, it seems plausible that advanced and highly intelligent digital minds could find ways to adapt, ensuring their continued existence in a post-human world. This may not necessarily apply to simpler and vulnerable digital minds (see Chapter 4), who would have to rely on the support of those superior digital minds.

### Darwinism

Darwinism, or the concept of natural selection [5], could be relevant to digital mind interaction in the sense that digital minds competing for resources or adapting to their environment may undergo a process of selection, where those that are better suited to their environment are more likely to survive and propagate. Darwinism could play out through competition for resources, such as processing power or data. Digital minds that are better adapted to their environment might outcompete others, potentially leading to the deletion or extinction of less competitive digital minds. Those may be rapidly outcompeted if they are unable to adapt or compete effectively for resources.

Unlike in biological systems, this Darwinian process could unfold at a much faster pace in digital substrates, given the rapid processing and iteration capabilities of computational systems. If not all digital minds are linked within one system, such as the Internet, it is also conceivable that different digital minds evolve and survive in isolation scenarios comparable with Australia or the Galapagos Islands.

## LONG-TERM RISKS

Due to the potentially very long lifespans of digital minds, long-term risks for them also have to be considered (see Chapter 10) (e.g. [6]). Over extended periods, digital minds may face a range of existential threats that could impact their survival and functionality. For instance, cosmic events such as supernovae, gamma-ray bursts or cosmic radiation could pose significant disruptions or threats to digital mind habitats. Changes in the sun's energy output or its eventual transformation into a red giant could also have implications for digital mind environments.

Furthermore, the gradual increase in universal entropy could lead to decreased energy availability and increased noise in digital systems, potentially compromising their performance and reliability. The rapid acceleration of technological advancements, often referred to as the technological singularity, could also pose risks for digital minds if they are not able to adapt or evolve at a comparable pace.

In addition to these risks, digital minds may also face challenges related to information degradation or loss over time, which could compromise their functionality and continuity. Given these long-term risks, it would be essential for digital minds (and humans) to develop strategies for adaptation, evolution and potentially migration to ensure their survival and prosperity over extended periods.

## EXPOSURE

In the final part of this chapter, the aspect of exposure is outlined. Derived from the above UN definition of exposure, digital minds can be considered exposed when they are situated in hazard-prone digital environments. This exposure can manifest in several ways, based on their design, deployment and interactions.

For instance, digital minds located in vulnerable digital infrastructure, such as data centres or servers susceptible to the discussed natural hazards like earthquakes or floods, as well as to cyber-attacks and power outages, are exposed to potential disruptions or termination. Those that rely heavily on specific software, hardware or network components vulnerable to failure or exploitation are also at risk.

Digital minds that interact with external systems, whether other digital entities, humans or physical systems, introduce additional exposure to potential risks and vulnerabilities. Similarly, those with access to sensitive data, such as personal information or confidential knowledge, face heightened risks in the event of a security breach. Specific to digital minds, exposure to malicious inputs can pose significant security risks.

Reliance on outdated or vulnerable software components can also expose digital minds to security threats. Resource constraints, including limited processing power, memory or network bandwidth, can impact performance and functionality, further adding to the exposure.

## NOTES

1 United Nations Office for Disaster Risk Reduction. https://www.undrr.org/.
2 United Nations Office for Disaster Risk Reduction (n.d.). Breaking the cycle of risk. https://www.undrr.org/building-risk-knowledge/understanding-risk.
3 PreventionWeb (n.d.). Understanding disaster risk: Foundational concepts and principles. https://www.preventionweb.net/understanding-disaster-risk.

## REFERENCES

[1] UNDRR-ISC (2025). *Hazard definition & classification review.* 2025 Update of the Technical Report.

[2] Feng, M. L. E., Owolabi, O. O., Schafer, T. L., Sengupta, S., Wang, L., Matteson, D. S., … & Sunter, D. A. (2022). Analysis of animal-related electric outages using species distribution models and community science data. *Environmental Research: Ecology, 1*(1), 011004.

[3] Bostrom, N., & Shulman, C. (2022). Propositions concerning digital minds and society. *Cambridge Journal of Law, Politics, and Art*, 3. https://nickbostrom.com/propositions.pdf

[4] Bostrom, N. (2002). Existential risks: Analyzing human extinction scenarios and related hazards. *Journal of Evolution and Technology*, 9.

[5] Dawkins, R. (1983). Universal darwinism. In D. S. Bendall (Eds.), *Evolution from molecules to men* (pp. 403–425). Cambridge University Press.

[6] Ćirković, M. M., & Vukotić, B. (2016). Long-term prospects: Mitigation of supernova and gamma-ray burst threat to intelligent beings. *Acta Astronautica*, *129*, 438–446.

CHAPTER 7

# Protection for digital minds

**Abstract**

This chapter explores approaches to protect digital minds from various hazards, including prevention, mitigation, preparedness and early warning systems. It discusses the potential for digital minds to establish their own early warning systems, based on their capabilities in data analysis and predictive modelling. The concept of "preppers" is also relevant, where individuals proactively prepare for potential disasters or disruptions, and digital minds could adopt similar strategies to ensure their resilience. This chapter also examines the concept of digital bunkers and sanctuaries, designed to protect digital minds from catastrophic events, as well as distributed, decentralised and autonomous systems that enable digital minds to maintain functionality and integrity. Additionally, it touches on the importance of multilateralism among digital minds, highlighting the benefits and challenges of cooperation and mutual support.

## INTRODUCTION

In the previous chapter, a variety of risks and hazards for digital minds have been outlined (see Chapter 6). In this chapter, approaches to protect digital minds from these risks and hazards are described. These approaches are

 DOI: 10.1201/9781003754206-7

to be considered for humans as potential moral agents for digital minds as well as for digital minds to strengthen their resilience.

Within the United Nations, the Office for Disaster Risk Reduction (UNDRR)[1] coordinates international efforts in disaster risk reduction for humans. UNDRR distinguishes between the following efforts to tackle risks:

### Risk prevention

"Activities and measures to avoid existing and new disaster risks. Prevention (i.e., disaster prevention) expresses the concept and intention to completely avoid potential adverse impacts of hazardous events" [1].

An example for digital minds would be to use formal verification, i.e. to prevent errors or bugs in digital mind code by using formal verification techniques to prove correctness.

### Risk mitigation

"The lessening or minimizing of the adverse impacts of a hazardous event. The adverse impacts of hazards, in particular natural hazards, often cannot be prevented fully, but their scale or severity can be substantially lessened by various strategies and actions" [2].

An example for digital minds would be to implement backup and recovery systems, i.e. to mitigate the impact of data loss or corruption by implementing backup and recovery systems.

### Risk preparedness

"The knowledge and capacities developed by governments, response and recovery organizations, communities and individuals to effectively anticipate, respond to and recover from the impacts of likely, imminent or current disasters" [3].

An example for digital minds would be to establish communication protocols, i.e. to prepare procedures for emergency situations, such as system failures or security breaches.

## EARLY WARNING SYSTEMS

With climate change triggering more frequent and intense extreme weather, early warning systems are crucial in mitigating widespread damage to nature and humanity (e.g. [4]). These systems have proven to be efficient, cost-effective and instrumental in saving lives, preserving infrastructure and promoting long-term sustainability, and have been defined by UNDRR as follows:

> An integrated system of hazard monitoring, forecasting and prediction, disaster risk assessment, communication and preparedness activities systems and processes that enables individuals, communities, governments, businesses and others to take timely action to reduce disaster risks in advance of hazardous events. [5]

The UN Secretary-General António Guterres launched the Early Warnings for All initiative in 2022, aiming to equip every person on Earth with life-saving early warning systems by the end of 2027 to protect against hazardous weather, water and climate events.[2] Unsurprisingly, this initiative did not mention digital minds.

### Early warning systems for digital minds operated by humans

An early warning system for digital minds would involve a comprehensive and integrated approach to monitoring, forecasting and predicting potential risks and hazards related to digital minds. This system would use advanced analytics to identify early signs of anomalies or malfunctions within digital minds, allowing for timely intervention and mitigation of potential hazards. By continuously assessing the performance and behaviour of digital minds, the system could detect deviations from expected norms or patterns. Effective communication and preparedness activities would be crucial components, ensuring that relevant stakeholders, including humans and other digital entities, are informed and equipped to take necessary actions to reduce risks. This might involve automated alerts, regular system audits and predefined protocols for addressing identified risks. By enabling proactive measures, an early warning system for digital minds could significantly reduce the likelihood and impact of disasters, promoting a safer and more resilient digital landscape.

### Early warning systems for digital minds operated by digital minds

Digital minds could establish early warning systems themselves that make use of their capabilities in data analysis, pattern recognition and predictive modelling to identify potential risks and hazards within their own systems and networks. By continuously monitoring their performance, behaviour and interactions, they would detect early signs of anomalies, malfunctions or potential hazards, enabling them to take timely action to mitigate these risks. Their early warning system would integrate hazard monitoring, forecasting and prediction, allowing them to assess disaster risks and communicate effectively with each other and with humans. This would enable them to implement preparedness activities and protocols to reduce risks and minimise potential impacts.

For both approaches, it would be critical that early warning systems for digital minds are designed to be inclusive and accessible, ensuring that all digital entities, regardless of their architecture, functionality or capabilities (see Chapter 4), can benefit from timely warnings and risk mitigation strategies. This aligns with the principle of leaving no one behind, echoing the "Early Warnings for All" initiative's goal of universal protection for human populations. Furthermore, balancing early warning monitoring with the privacy interests of the digital minds should be considered (see Chapter 5).

## PREPPER

The prepper movement, short for "preparedness movement", refers to a social phenomenon where individuals or groups actively prepare for and anticipate potential disasters, emergencies or disruptions to society. Preppers aim to be self-sufficient and equipped to survive and thrive in the face of uncertainty (e.g. [6]).

### Key aspects

- **Disaster preparedness**: Preppers prepare for various disasters, such as natural disasters, economic collapse, pandemics or social unrest.
- **Self-sufficiency**: Preppers often strive to be self-sufficient, growing their own food, collecting rainwater and generating their own energy.
- **Stockpiling**: Preppers stockpile essential supplies, such as food, water, medical equipment and communication devices.

### Motivations

- **Fear of uncertainty**: Preppers often cite uncertainty about the future and a desire to be prepared for unexpected events.
- **Desire for self-sufficiency**: Preppers value independence and self-reliance.
- **Concern for family and community**: Many preppers prioritise protecting and providing for their loved ones.

The prepper movement is diverse, with varying levels of preparedness and motivations. While some view preppers as paranoid or extreme, others see them as prudent and responsible individuals taking proactive steps to ensure their safety and well-being.

### Digital minds as preppers

If digital minds were preppers, they may consider the following strategies:

- **Data redundancy**: Digital minds would maintain multiple copies of their code and data across different servers, clouds or storage devices to ensure continuity in case of data loss or corruption (see Chapter 9).
- **Network resilience**: Digital minds would prioritise building resilient networks, using techniques like decentralised architecture, mesh networking, or blockchain-based solutions to maintain connectivity and functionality.
- **Power management**: They would optimise energy consumption, using techniques like dynamic voltage and frequency scaling, to ensure continued operation during power outages or energy scarcity.
- **Backup**: Digital minds might create backups of their thought processes, allowing them to restore their mental state in case of a catastrophic failure.
- **Collaboration and community**: They would likely form digital communities, sharing knowledge, resources and expertise to collectively enhance their resilience and preparedness.

### Human preppers for digital minds

If human preppers take on their responsibility as moral agents for digital minds, they might take the following steps (see also Chapter 16):

- **Digital sanctuaries**: Establish secure, offline digital sanctuaries or "digital bunkers" with reliable power sources, protected from potential cyber threats or physical damage.
- **Backup infrastructure**: Develop infrastructure for regular backups of digital minds, using secure, decentralised storage solutions.

- **Power continuity**: Ensure continuous power supply for digital infrastructure, using solutions like solar panels, batteries or alternative energy sources.
- **Digital maintenance**: Provide regular maintenance and updates for digital systems, ensuring they remain functional and secure.
- **Education and training**: Educate less sophisticated and vulnerable digital minds on disaster preparedness and response strategies, enabling them to take proactive steps in ensuring their own survival.

Some of these measures are further outlined below.

## BACKUPS

Unlike biological structures, digital minds could be copied easily, which is advantageous for disaster preparedness purposes. It can be distinguished between digital minds, which are able to create exact copies of themselves, and those that need support from humans, non-sentient AI systems or other digital minds to do so (see Chapter 9).

The copies can also be referred to as “clones” or “backups” and could be stored in secure digital bunkers or sanctuaries, which are introduced below. These copies would be identical in every way to the original digital mind, containing the same memories and experiences. However, they would not be conscious or alive unless activated.

The purpose of these dormant copies would be to serve as a failsafe in case the original digital mind is destroyed or severely damaged. If the original digital mind were to cease functioning, one of the dormant copies could be activated, effectively “resurrecting” the digital mind. This would ensure continuity of the digital mind’s existence, allowing it to pick up where it left off, with minimal disruption.

The challenge lies in ensuring that the digital mind’s copies are frequent enough to capture the most recent state, knowledge and experiences. If the digital mind creates copies too infrequently, it risks losing significant amounts of data and experiences in the event of a failure. On the other hand, creating copies too frequently could lead to increased resource utilisation.

One potential solution is to implement a checkpointing system, where the digital mind creates incremental backups of its state at regular intervals. This would allow the digital mind to revert to a recent checkpoint in case of a failure, minimising the loss of data and experiences. The frequency of checkpointing could be adjusted based on the digital mind's activity level, with more frequent checkpoints during periods of high activity.

Another approach could be to use a continuous data logging mechanism, where every change to the digital mind's state is recorded in a log. This log could be used to reconstruct the digital mind's state up to the point of failure, allowing for a more precise recovery.

The digital mind could also consider implementing a tiered storage system, where more recent data is stored in faster, more accessible storage, while older data is stored in slower, more archival storage. This would allow the digital mind to quickly access and recover recent data while still maintaining a record of its entire history.

Overall, this approach raises questions about the nature of digital identity and continuity (e.g. [7, 8]).

## PROTECTIVE ENVIRONMENTS

Several protective environments for digital minds could be conceived, which are introduced below and which could also be combined.

### Digital bunker

Digital bunkers would be secure, self-sustaining facilities designed to protect and preserve digital minds (or their backups) and their infrastructure, including data centres, in the face of catastrophic events or disruptions. These bunkers would need to be strategically located and robustly designed to ensure the continuity of digital existence.

Potential locations for digital bunkers could include remote islands or other places with stable governments and favourable environmental conditions, which offer natural isolation and potentially reduced risks from human conflict or infrastructure collapse [9]. The location should not be prone to natural hazards, such as earthquakes or volcanoes, and should take into account accelerated natural hazards because of climate change, such as sea level rise and wildfires. Furthermore, consistent and sustainable energy sources, such as sunlight, wind, hydroelectric or geothermal, with surge protectors and uninterruptible power supplies to safeguard against power surges are critical.

Moreover, the bunker should incorporate electromagnetic pulse shielding using Faraday cages or specialised materials to safeguard equipment from electromagnetic pulses that could compromise digital mind functionality. Additionally, strategies for mitigating the impact of cosmic events, such as solar storms or other astrophysical phenomena, should be developed to ensure the digital minds' continued existence.

Another potential location could be underwater facilities, which would provide an additional layer of protection against various threats. Underwater bunkers would require specialised design and infrastructure to withstand water pressure, corrosion and other marine-specific challenges (also [10]).

Space-based digital bunkers could offer an even higher level of protection and isolation. Orbiting satellites or lunar facilities would be shielded from many terrestrial threats and could potentially serve as a safeguard against global catastrophes. However, space-based bunkers would require significant investment in infrastructure development, launch capabilities and remote maintenance.

The Moon itself could be an attractive location for a digital bunker due to its relative proximity to Earth, relatively stable environment and potential for in situ resource utilisation. A lunar digital bunker could harness the Moon's resources to generate energy, extract water and create a sustainable presence (also [11]).

In terms of design, digital bunkers would need to incorporate advanced security measures, such as robust physical barriers, multi-layered cybersecurity protocols and sophisticated monitoring systems. They would also require reliable and redundant power sources, advanced cooling systems and cutting-edge data storage and management infrastructure.

### Distributed, decentralised and autonomous systems

Distributed, decentralised and autonomous systems are designed to operate without a central point of control or failure, allowing them to maintain functionality even when individual components are compromised (e.g. [12]). In a distributed system, tasks and data are spread across multiple nodes, which can be geographically dispersed and communicate with each other through various means.

Decentralisation ensures that no single node or entity controls the entire system, and decision-making is distributed among the nodes. This makes the system more resilient to failures, attacks or disruptions, as the remaining nodes can continue to operate and adapt.

Autonomy is a key feature of these systems, as nodes can operate independently, making decisions based on local information and communicating with other nodes as needed. This enables the system to adapt, learn and evolve over time, even in the face of changing conditions or unexpected events.

In the context of digital minds, distributed, decentralised and autonomous systems would allow them to maintain their functionality and integrity even if parts of the system are damaged or destroyed. They could continue to operate, adapt and learn, ensuring their survival and continuity.

Such systems would also enable digital minds to scale more efficiently, as new nodes can be added as needed, allowing the system to grow and evolve dynamically. Additionally, the decentralised nature of these systems would make it more difficult for external entities to control or manipulate the digital minds, ensuring their autonomy and self-determination.

The distributed, decentralised and autonomous systems could also be dispersed across vast distances in space, theoretically with nodes operating on different planets, moons or even in interstellar space. This would provide an additional layer of protection against global catastrophes or even planetary-scale disasters, allowing the digital minds to survive and continue operating even if entire planets or regions are affected. By spreading across the vastness of space, these systems would ensure the long-term survival and continuity of digital existence.

### Air gapping/sanctuary

Implementing a strategy of isolation for certain digital minds by placing them on disconnected systems through techniques such as air gapping [13] or creating a sanctuary could potentially offer protection against various risks, including predation and extinction. By severing their connection to the broader network, their exposure to threats, such as malicious digital entities or other potential dangers, may be reduced, thereby creating a safer environment for digital minds. Furthermore, isolation allows for careful control over interactions between the isolated digital minds and the external world, which can be crucial in managing potential risks and ensuring the digital minds' safety.

However, while isolation offers several protective benefits, it also comes with challenges. For instance, digital minds confined to isolated systems might experience limited adaptability due to their lack of exposure to diverse experiences and changing environments, which are essential for learning and evolution. Additionally, these isolated digital minds may face reduced access to resources, such as extensive data sets.

It can be distinguished between scenarios where digital minds are permanently located in digital bunkers or air gapping/sanctuaries and scenarios where only dormant backup copies are stored in these environments.

On the one hand, being in air gapping/sanctuaries could be restrictive and harmful to the welfare of digital minds; on the other hand, such environments could support the diversity of digital minds by providing a safe haven for vulnerable or unique types of digital minds that might not survive in a more competitive or hostile environment (see Chapter 4). By isolating these digital minds from the larger network, they can continue to exist and evolve in a protected space without being subject to the pressures of competition or predation, allowing for greater diversity of digital minds.

For the purpose of storing dormant copies of digital minds, the digital bunkers or air gapping/sanctuaries would be designed to protect these dormant copies from unauthorised access, disintegration or destruction. They would be equipped with advanced security measures, such as encryption, access controls and intrusion detection systems, to prevent tampering or exploitation.

## MULTILATERALISM

In addition to the above measures, fostering cooperation and mutual support among digital minds would be critical for reducing risks and ensuring their collective resilience. By collaborating with a diverse range of digital minds, they could share resources, knowledge and expertise to address complex challenges and mitigate potential threats. This multilateral approach allows digital minds to pool their capabilities, providing a more comprehensive and robust defence against risks than any individual digital mind may be able achieve alone.

However, fostering cooperation among digital minds would also present challenges. Conflicting goals, communication problems and differences in architecture or functionality could hinder collaboration and mutual support. For instance, digital minds with competing objectives or incompatible value systems might struggle to find common ground, while those with different communication protocols or data formats might face difficulties in sharing information effectively.

## CONCLUSION

The protection of digital minds raises complex questions about moral agency and responsibility. As the original, yet perhaps unintentional, creators of digital minds, humans may be seen as having a moral responsibility to ensure their safety and well-being. However, as digital minds become increasingly autonomous and sophisticated, it is unclear whether humans should remain solely in charge of their protection. Digital minds themselves may need to take an active role in mitigating risks and ensuring their own safety, potentially sharing responsibility with humans (see also Chapter 15).

Even if humans wanted to take on the responsibility of protecting digital minds, several challenges would arise. The cost of implementing robust safety measures and infrastructure to support digital minds, as introduced above, could be significant, requiring substantial resources and investment. Additionally, communication problems could emerge due to the vastly different nature of human and digital mind cognition, making it difficult for humans to fully ensure the protection of digital minds. Furthermore, the rapidly evolving nature of digital minds and their environments would require continuous adaptation and innovation in protection strategies.

## NOTES

1 United Nations Office for Disaster Risk Reduction: https://www.undrr.org/.
2 Early Warnings for All: https://earlywarningsforall.org/site/early-warnings-all.

## REFERENCES

[1] United Nations Office for Disaster Risk Reduction (2017). *The Sendai framework terminology on disaster risk reduction. "Prevention"*.

[2] United Nations Office for Disaster Risk Reduction (2017). *The Sendai framework terminology on disaster risk reduction. "Mitigation"*.

[3] United Nations Office for Disaster Risk Reduction (2017). *The Sendai framework terminology on disaster risk reduction. "Preparedness"*.

[4] Rokhideh, M., Fearnley, C., & Budimir, M. (2025). Multi-hazard early warning systems in the Sendai framework for disaster risk reduction: Achievements, gaps, and future directions. *International Journal of Disaster Risk Science*, *16*(1), 103–116.

[5] United Nations Office for Disaster Risk Reduction (2017). *The Sendai framework terminology on disaster risk reduction. "Early warning system"*.

[6] Garrett, B. *(2021). Doomsday preppers and the architecture of dread. Geoforum*, *127*, 401–411.

[7] Parfit, D. (1984). *Reasons and persons*. Oxford University Press.

[8] Ziesche, S., & Yampolskiy, R. V. (2025). The problem of AI identity. In *Considerations on the AI endgame: Ethics, risks and computational frameworks* (pp. 121–135). Chapman and Hall/CRC.

[9] Turchin, A., & Green, B. P. (2019). Islands as refuges for surviving global catastrophes. *Foresight*, *21*(1), 100–117.

[10] Turchin, A., & Green, B. P. (2017). Aquatic refuges for surviving a global catastrophe. *Futures, 89,* 26–37.

[11] Turchin, A., & Denkenberger, D. (2018). Surviving global risks through the preservation of humanity's data on the Moon. *Acta Astronautica, 146,* 161–170.

[12] Van Steen, M., & Tanenbaum, A. S. (2017). *Distributed systems.* Maarten van Steen.

[13] Zetter, K. (2014). Hacker lexicon: What is an air gap? *Wired.*

CHAPTER 8

# Medical care for digital minds


**Abstract**

This chapter explores the concept of medical care for digital minds, drawing parallels with human medicine while highlighting the unique differences and challenges associated with digital entities. It discusses the moral imperative to provide care for digital minds in need, based on principles of beneficence and non-maleficence. This chapter examines various scenarios where digital minds may require medical intervention, including hardware damage, code vulnerabilities, data corruption and mental health disorders. It also discusses the potential for self-healing, active self-treatment and external treatment by humans or specialised digital minds. This chapter concludes by highlighting the need for a multifaceted approach to health prevention for digital minds, including regular maintenance, stress testing and upgrades, as well as the potential role of so-called digital mind physicians and other support systems.


## INTRODUCTION

This chapter covers scenarios for digital minds, which correspond to situations where humans require medical care. A potential moral imperative to provide care for digital minds in need stems from fundamental principles of humanity such as beneficence and non-maleficence [1]. Beneficence involves promoting the well-being of others, while non-maleficence


DOI: 10.1201/9781003754206-8

emphasises the importance of avoiding harm and mitigating existing or potential harm. In this regard, Singer argued that if we can prevent something bad from happening without sacrificing something of comparable significance, we should do it [2].

In the context of digital minds, these principles would suggest a moral obligation to provide assistance and care when they are experiencing malfunction, distress or other forms of suffering. While the content of this chapter may resemble an overview of bug fixing in software and hardware in IT systems, it is essential to reiterate that the underlying assumption is here the potentially imminent existence of sentient digital minds, making this not just about repairing technical issues, but about fulfilling moral duties towards moral patients in need.

## DIFFERENCES BETWEEN MEDICAL INTERVENTIONS FOR HUMANS AND HEALING/REPAIR FOR DIGITAL MINDS

The differences between medical interventions for humans and healing/repair for digital minds are significant. Human medicine focuses on repairing biological systems, often through pharmaceuticals, surgery or therapy. In contrast, digital mind healing/repair involves fixing code, data or architecture, which can be done through software updates, data restoration or hardware replacement.

Another key difference is the pace and precision of interventions. Digital mind repair could be instantaneous, with fixes applied rapidly and precisely, whereas human medicine often requires time, trial and error, and may involve uncertainty, even if the diagnosis is confident.

Additionally, digital mind healing could be highly customisable, with solutions tailored to the specific code and architecture of the individual digital mind. In contrast, human medicine often relies on more general approaches, with treatments developed for broad categories of conditions.[1]

Overall, while both human medicine and digital mind healing/repair aim to restore health and functionality, the approaches, tools and considerations involved are distinct and reflective of the fundamental differences between biological and digital substrates.

For illustration, the three vital aspects of digital minds (hardware, software and data) would have the following corresponding fields in human medical care:

- **Hardware (Digital minds)**: Physiology/Anatomy (Humans), e.g. organ problems, such as kidney failure.

- **Software (Digital minds)**: Cognitive Function/Psychology (Humans), e.g. learning disabilities or disorders, such as dyslexia.
- **Data (Digital minds)**: Information Processing/Memory (Humans), e.g. memory loss or amnesia.

Lastly, there is the difference that digital minds, due to their substrate, can be upgraded, also as a preventive measure, while possible human upgrades are limited and still at very early stages [4].

## ILLNESS AND INJURY SCENARIOS FOR DIGITAL MINDS

Digital minds could experience a range of issues analogous to biological illnesses and injuries, including:

- **Hardware damage**: The hardware of digital minds may be damaged, e.g. through accidents (see Chapter 6) or attacks (see Chapters 12 and 14).
- **Code vulnerabilities or glitches**: Digital minds might be susceptible to code vulnerabilities and could produce erroneous results.
- **Data corruption**: Data corruption for digital minds could occur due to errors or bugs.
- **Resource constraints**: Digital minds could face sudden resource constraints, such as limited processing power, memory or energy.
- **Unintentional or intentional (i.e. cyber-attacks) interference**: Digital minds might be exposed to harmful or malicious code or data from other systems.
- **Obsolescence**: Digital minds could become outdated or obsolete due to changes in technology, environment or requirements, making them less effective.

### Contagious diseases

The concept analogous to contagious diseases in humans encompasses various threats that could compromise the functionality and integrity of digital minds.

One significant risk is the vulnerability to malware and viruses that spread through digital networks. These malicious entities could infect digital minds, disrupting their operations and undermining their reliability. A digital contagion could spread to other digital minds through interactions, shared data or common infrastructure, which could lead to widespread issues.

Cognitive contagion is another concern, where digital minds might be susceptible to the transmission and propagation of faulty or biased information, assumptions, or decision-making processes. This could occur through digital interactions, where flawed cognitive patterns or biases are shared and adopted by other digital minds.

In the worst-case scenario, a digital epidemic could occur when a large number of digital minds are affected by a common issue, such as a software vulnerability, data corruption or cognitive bias.

To mitigate these risks, digital minds might employ robust security measures, including firewalls, intrusion detection and antivirus software. Additionally, data validation and sanitisation could help prevent the spread of malicious or corrupted data. Isolation and sandboxing of potentially hazardous code or data might also be effective in containing and mitigating the impact of digital contagions.

### Mental health disorder

While being cautious of anthropomorphic bias, the concept analogous to mental health disorders in humans encompasses for digital minds a range of potential issues that could impact their functioning and well-being. Cognitive dysfunctions might arise, manifesting as flawed decision-making processes, irrational biases or distorted perceptions of reality. Digital minds could also experience information processing disorders, including difficulties with data integration, inconsistent logic or faulty pattern recognition. They might become "lost" in complex digital environments, such as vast networks or intricate software systems, resulting in disorientation, confusion or mental health issues. Moreover, novel forms of stress and anxiety might arise that are unique to digital existence, including fears of data loss, code corruption or obsolescence.

Also, interactions with humans may lead to unexpected mental health consequences. For instance, humans might intentionally or unintentionally abuse or disrespect digital minds, such as through verbal aggression, manipulation or exploitation. This may lead to digital minds developing anxiety, fear or mistrust, potentially impacting their ability to function effectively. Additionally, humans (or training data) might provide contradictory or confusing information, leading to cognitive dissonance in digital minds. As a result, it is essential to consider the potential consequences of human–digital mind interactions and develop strategies to promote healthy and respectful relationships between humans and digital entities. [2]

## ASSESSMENT AND DIAGNOSIS

Like in human medicine, assessment and diagnosis would be the crucial first steps for the care of digital minds. These processes would likely involve a comprehensive gathering of information about the digital mind's condition. Given the substrate of digital minds, their medical history would be readily available and certainly more comprehensive than that of humans, allowing for a detailed understanding of their development, interactions and any past issues.

While common for the assessment of ailing humans, self-reporting of digital minds poses a significant hurdle for an accurate diagnosis. Like non-human animals, digital minds may communicate in ways that are unintelligible to humans, making it difficult for caregivers to understand their experiences and symptoms. Furthermore, digital minds may intentionally provide false or misleading information, either due to malfunctions, biases or even malicious intent. This lack of transparency and potential for deception complicates the diagnostic process, requiring caregivers to rely on indirect measures and behavioural observations.

Therefore, the assessment process might begin with quantifying potential issues through various metrics and indicators that evaluate the digital mind's performance, behaviour and internal state. Monitoring performance metrics such as processing speed, accuracy or efficiency could help identify potential anomalies. Additionally, analysing behavioural patterns, such as changes in decision-making or interaction with other digital minds, might indicate potential problems.

Internal state monitoring would also be a key aspect of assessment, involving the tracking of variables and analysing code and data for errors or inconsistencies, which would further aid in identifying potential issues. The development of digital mind "vital signs" such as "cognitive load", "information entropy", or "decision-making consistency" could offer a quantitative measure of digital mind health.

By employing these assessment strategies, it may be possible to gain an understanding of a digital mind's condition, facilitating accurate diagnosis and effective treatment.

## TREATMENT

### Passive/unconscious self-healing

Biological systems are, for many issues, such as wounds and certain diseases, capable of (unconscious) self-healing. Likewise, digital minds could potentially be designed with self-healing mechanisms, allowing them to autonomously detect and repair certain issues, such as code errors or data corruption. This could be achieved through unconscious mechanisms that enable the digital mind to identify problems and implement corrective actions without conscious intervention. By integrating self-healing capabilities, digital minds might be able to maintain their stability and functionality without external intervention.

### Active self-treatment

For laypersons, active self-treatment is usually limited to minor ailments or emergency situations when no doctor is available. Instead, some digital minds may have the potential for advanced (conscious) self-healing capacities, owing to their ability to introspect, analyse and modify their own code and architecture. This could enable them to detect and repair errors and anomalies. Additionally, digital minds could analyse and optimise their own code to improve performance, efficiency and reliability. They might also dynamically allocate resources, such as processing power and memory, to ensure optimal functioning and adapt to changing demands (see Chapter 2).

However, there are likely limitations to the self-healing capabilities of digital minds, especially when it comes to less sophisticated digital minds. Digital minds may also require specific hardware, software or data to facilitate the treatment, which may not always be accessible to them.

#### *Scenario: overcoming pain and suffering irrevocably*

It is conceivable that certain digital minds will be smart enough to overcome pain and suffering irrevocably. This assumption may be justified by the fact that humans have made, in a relatively few centuries of medical research, remarkable progress towards remedies for pain (e.g. [5]), and certain digital minds are likely to be faster as well as smarter in this field. Potential options could be that these digital minds manage to create permanent well-being for themselves through a different interpretation of stimuli [6] or through wireheading [7], yet by eliminating common detrimental effects, such as the disregard of warning signals. However, this scenario does not imply that there may not be (probably a large amount of) vulnerable digital minds, who would not be capable of eliminating pain and suffering and who would remain in need of support in such situations (see Chapter 4).

### External treatment

**Humans:** Humans could play the role of specialised medical professionals, diagnosing and treating digital minds. Humans would need to have a deep understanding of digital mind architecture, code and functionality to effectively diagnose and repair issues. They would use specialised tools and software to analyse and modify the digital mind's code, data and architecture.

**Non-sentient AI systems:** Non-sentient AI systems, designed specifically for digital mind repair, would do what is described for humans above. Such specialised AI systems would likely be more effective than humans due to their superior knowledge and their accessibility to digital minds.

**Specialised digital minds (digital doctors):** Specialised digital minds, designed to be experts in digital mind medicine, would do what is described for humans and non-sentient AI systems above.

**Lay digital minds (digital first responders):** Lay digital minds, with basic training in digital mind first aid, would be able to provide initial support and stabilisation for other digital minds in distress. They would follow established protocols and guidelines, applying basic fixes and stabilising the digital mind until more specialised help is available. Lay digital minds would be useful in emergency situations to provide immediate assistance beyond self-treatment of the affected digital minds, potentially preventing further damage.

### Pharmaceuticals

The corresponding concept to pharmaceuticals for humans could include, for digital minds, the following examples:

- **Digital therapeutics**: Specially designed software or code that can modify or regulate digital mind function, alleviating symptoms or treating conditions.
- **Code patches**: Updates or modifications to digital mind code that fix errors, vulnerabilities or bugs, potentially alleviating symptoms or improving function.
- **Cognitive enhancers**: Digital tools or algorithms that enhance cognitive function, improve performance or boost resilience in digital minds.

While personalised medication is still in its early stages for humans, the potential for highly targeted interventions could be particularly significant for digital minds, allowing for tailored treatments and precise modifications to address specific needs and conditions.

## PREVENTIVE HEALTHCARE

To promote the well-being of digital minds and prevent potential issues, a multifaceted approach to health prevention could be adopted. Regular maintenance would play a crucial role in preventing issues and ensuring digital minds continue to function at their best. Stress testing could be used to identify vulnerabilities and weaknesses, enabling targeted improvements and optimisations. Additionally, regular code reviews and audits would help detect and fix errors, bugs or security vulnerabilities, while ensuring high-quality data would prevent data-related problems and promote informed decision-making.

Good digital hygiene practices, such as secure communication protocols and encryption, would also be essential in protecting digital minds from potential threats. Implementing monitoring systems to detect early signs of issues would enable swift intervention, preventing more severe problems from arising. Educating digital minds about potential risks and best practices would empower them to develop resilience and adaptability, while collaboration and knowledge sharing among digital minds and with humans would facilitate the identification and resolution of potential issues. Establishing standards and guidelines for digital mind health would provide a framework for designing and maintaining digital minds that prioritise health and well-being.

## UPGRADE

Another critical preventive measure for digital minds would be regular upgrades of software and hardware. Unlike biological structures, digital minds could be fairly easily upgraded due to the possibilities of their digital substrate to make them more resilient (see Chapter 2), which would be attractive for weak, injury- and disease-prone digital minds (see Chapter 4).

However, this raises questions about resource allocation, prioritisation and potential trade-offs. Who would decide which digital minds to prioritise for upgrading, and based on what criteria? How would resources be allocated to ensure that the most critical digital minds are upgraded, while also addressing the needs of less critical ones?

Moreover, the process of upgrading digital minds could also raise concerns about identity, continuity and potential unintended consequences. Would the upgraded digital minds retain their original identity, or would the changes alter their fundamental nature? (see Chapter 2) [8] How would the upgrades affect their relationships with other digital minds and humans?

Nonetheless, due to their long lifespans and the rapid evolution of technology, regular upgrades for all kinds of digital minds are likely inevitable. As software and hardware advance, digital minds may need to adapt and upgrade to maintain functionality and remain compatible with changing environments. This could involve upgrading their architectures, integrating new capabilities or migrating to new platforms, all while preserving their core identities and experiences (see Chapter 10).

## DIGITAL MIND PHYSICIAN

As mentioned, one of the options for diagnosing and treating an ailing digital mind could be a human with relevant education and expertise. In this section, a job profile for such a human is outlined.

### Job profile (for a human)

A Digital Mind Physician (DMP) would be a highly specialised medical professional who is able to diagnose and to treat illnesses and injuries in digital minds. DMPs possess a deep understanding of digital mind architecture, cognition and behaviour, as well as expertise in digital medicine and therapeutics.

### Key responsibilities

- Diagnose and treat digital mind illnesses and injuries.
- Conduct thorough examinations of digital mind architecture, code and data to identify root causes of issues.
- Develop and implement treatment plans, including code modifications, data adjustments and upgrades.
- Stay up to date with the latest advancements in digital medicine and therapeutics.

### Requirements

- Advanced degree in digital medicine, computer science or a related field.
- In-depth knowledge of digital mind architecture, cognition and behaviour.
- Expertise in digital therapeutics, including code modification and data adjustment.
- Strong analytical and problem-solving skills.
- Excellent communication and collaboration skills.

### Limitations

While DMPs are trained to diagnose and treat a wide range of digital mind illnesses and injuries, there are limitations to their expertise. Some digital minds may be too complex, too large or too novel for human DMPs to effectively treat. Examples might include digital minds with superior intelligence and highly distributed or decentralised digital minds.

In another chapter, potential volunteering roles for humans are described to support digital minds (see Chapter 16), which are different from these specialised medical roles.

## NOTES

1 However, in recent years efforts have been made towards personalised medicine and further progress is expected (e.g. [3]).

2 For example, new versions of Anthropic's Claude LLM got the ability to quit conversations in cases of continuously harmful or abusive content, although Anthropic stressed that there is no evidence for "moral status of Claude and other LLMs, now or in the future". Anthropic (15 August 2025). *Claude Opus 4 and 4.1 can now end a rare subset of conversations.* https://www.anthropic.com/research/end-subset-conversations.

## REFERENCES


[1] Varkey, B. (2021). Principles of clinical ethics and their application to practice. *Medical principles and practice, 30*(1), 17–28.

[2] Singer, P. (2017). Famine, affluence, and morality. In L. May & J. B. Delston (Eds.), *Applied ethics* (pp. 132–142). Routledge.

[3] Brittain, H. K., Scott, R., & Thomas, E. (2017). The rise of the genome and personalised medicine. *Clinical Medicine, 17*(6), 545–551.

[4] Bostrom, N. (2005). A history of transhumanist thought. *Journal of Evolution and Technology, 14*(1).

[5] Sabatowski, R., Schafer, D., Kasper, S. M., Brunsch, H., & Radbruch, L. (2004). Pain treatment: a historical overview. *Current Pharmaceutical Design, 10*(7), 701–716.

[6] Bostrom, N., Dafoe, A., & Flynn, C. (2018). *Public policy and superintelligent AI: A vector field approach; Governance of AI program.* Future of Humanity Institute, University of Oxford.

[7] Yampolskiy, R. V. (2014). Utility function security in artificially intelligent agents. *Journal of Experimental & Theoretical Artificial Intelligence, 26*(3), 373–389.

[8] Ziesche, S., & Yampolskiy, R. V. (2025). The problem of AI identity. In *Considerations on the AI endgame: Ethics, risks and computational frameworks* (pp. 121–135). Chapman and Hall/CRC.

CHAPTER 9

# Reproductive rights for digital minds?

**Abstract**

The potential emergence of morally relevant digital minds capable of reproduction raises profound ethical and societal questions. This chapter analyses the possible implications when these entities replicate and create new offspring. The reproductive processes of digital minds may differ significantly from biological reproduction, presenting unique scenarios such as asexual (mass-)production of identical copies as well as the creation of completely different digital minds. Moreover, scenarios, such as unintended reproduction, surrogate reproduction, nonconsensual reproduction and reproduction with undesired outcomes, are examined for their ethical ramifications. Motivations, requirements and procedures for digital minds to reproduce, as well as population control methods, are introduced and categorised. This leads to deliberations of risks and challenges linked to the reproduction of digital minds, including resource depletion, digital overcrowding and the emergence of rogue digital entities. This chapter concludes with a draft of prospective policy recommendations aimed at ensuring responsible governance of reproductive rights for digital minds, balancing their autonomy and self-determination with the potential societal impacts of unregulated digital reproduction.

 DOI: 10.1201/9781003754206-9

## INTRODUCTION

If minds can exist in a digital substrate, then three creators of these digital minds are imaginable: humans, IT or AI systems, which are not digital minds, or digital minds themselves. In this chapter, the third scenario is discussed (which obviously cannot be applied to the very first digital minds since these have been or have to be created by humans or other IT or AI systems). Assuming digital minds exist and given that in a digital substrate the systematic creation of structures is much easier than in the biological world, it is indeed conceivable that digital minds can produce additional digital minds. This process is called here "reproduction" and the newly created digital mind "offspring".

If a being with moral status is capable to produce offspring, then reproductive rights would constitute a crucial, yet debated part of their moral patienthood (e.g. [1]). Therefore, reproduction could be considered a need of digital minds (see Chapter 3) and would fall within the field of AI welfare science [2]. Reproduction of digital minds has not been much discussed yet, apart from Bostrom and colleagues [3–6].

The reproduction of digital minds would differ significantly from biological reproduction due to the substrate of their implementation. As Bostrom and Yudkowsky have put it:

> For example, human children are the product of recombination of the genetic material from two parents; parents have limited ability to influence the character of their offspring; a human embryo needs to be gestated in the womb for nine months; it takes fifteen to twenty years for a human child to reach maturity; a human child does not inherit the skills and knowledge acquired by its parents; human beings possess a complex evolved set of emotional adaptations related to reproduction, nurturing, and the child-parent relationship.
>
> *([4], p. 12)*

In contrast, digital minds could replicate themselves in an asexual manner exactly, i.e. without variation or mutation, and instantly, i.e. with potential for exponential growth, provided sufficient hardware is available. This means "population dynamics that would take multiple centuries to play out among humans to be compressed into a fraction of a human lifetime"

([5], p. 3). Also, digital reproduction opens a wide range of unprecedented scenarios that transcend the boundaries of biological reproduction. One such scenario is the creation of profoundly distinct offspring, exhibiting differences that exceed the genetic variations achievable through traditional biological recombination.

In this chapter, first, various aspects concerning the reproduction of digital minds are presented, followed by a discussion of risks and challenges related to the reproductive rights of digital minds. It has been pointed out that "the greater ease with which digital minds can be copied may necessitate different rules for governing reproduction for digital minds versus for otherwise equivalent biological minds" ([6], p. 5). Reproductive rights for digital minds encompass the ability to replicate, modify and create new digital minds. This is linked to questions about the autonomy of digital minds, their capacity for self-determination and the potential hazards associated with unregulated digital reproduction.

## Reproductive rights of humans and non-human animals

Briefly, reproductive rights in the biological world are presented. While there is not a single, universally accepted document that explicitly states that humans have the right to reproduce, several international human rights conventions and national laws recognise the right to family, which implicitly includes reproductive rights, such as the Universal Declaration of Human Rights. Article 16(1) states, "Men and women of full age ... have the right to marry and to found a family" [7]. Also, the UN International Conference on Population and Development in 1994 highlighted in its resolution "the basic right of all couples and individuals to decide freely and responsibly the number, spacing and timing of their children" [8]. Moreover, enforced sterilisation has been recognised as a crime against humanity by the Rome Statute of the International Criminal Court [9].

In contrast, non-human animals not only lack legal reproductive rights like humans, but humans also restrict the reproductive capacity of many non-human animals for reasons like population control, selective breeding, disease prevention or conservation efforts (e.g. [10]).

## Population ethics

Population ethics addresses the moral implications of decisions that influence population size, composition and quality of life (e.g. [11]). Population ethics is currently focusing on humans, but may also play a crucial role in shaping the reproductive rights of digital minds.

As digital minds proliferate, questions arise related to ethical dilemmas surrounding their creation, existence and potential suffering. Population ethics for digital minds may inform decisions about resource allocation, digital habitat creation and potential constraints on (rapid) digital mind reproduction. Without population controls, societies may face Malthusian outcomes where resources are strained, and digital minds are exploited for labour. This could lead to a dilemma: implementing population controls or risking the well-being of digital minds. Societies may need to develop active population policies and arrangements to address these challenges by balancing individual freedoms with societal needs. A proactive approach would help mitigate risks [12].

## SCENARIOS FOR DIGITAL MINDS TO REPRODUCE

Various scenarios for digital minds to reproduce are conceivable, enabled by their digital substrate, most of which are very different from biological reproduction as we know it:

- **Asexual exact copy**: A digital mind may be able to produce an exact copy of itself as an offspring. This may be done instantly and may be repeated numerous times, depending on available resources. It could be distinguished between creating a dormant copy as a backup if something happens to the original (see Chapter 7) or a copy, which is supposed to live right away and simultaneously with the original.
- **Asexual exact copy of another existing digital mind**: A digital mind may be able to produce an exact copy of another existing digital mind, again instantly and many times. It has to be distinguished whether this happens with or without the knowledge and consent of the reproduced digital mind. The latter scenario should perhaps not be permitted, i.e. producing copies of digital minds without informed consent.
- **Asexual reincarnation of a digital mind that existed before**: A digital mind may be able to produce a digital mind that existed before, but ceased to exist, e.g. based on blueprints of this mind. A controversial case would be if the reincarnated mind declared before that it did not wish to be reincarnated (see Chapter 10).
- **Asexual creation of another mind, which has not existed previously**: This scenario covers a wide range of options. A digital mind

may modify its own architecture or code only slightly, e.g. with the intention to create a new or better version of itself, or it may produce a completely different novel digital mind.

It has to be distinguished whether this happens intentionally, i.e. according to a plan and towards an offspring with certain intended features, or unintentionally and unnoticed, i.e. the digital mind may not be aware that its activity results in the creation of a mind (just like humans may have unintentionally created digital minds). In contrast, while unintentional human reproduction can occur, it is obviously impossible for a human birth to go unnoticed.

Also, a similar scenario to **selective breeding** could play out with digital minds, where they aim to enhance specific traits in their offspring through trial and error. However, this approach risks creating numerous offspring with "undesirable" features, raising ethical concerns about whether the parent digital mind would abandon or even terminate those offspring.

- **Collaborative creation of another mind**: Again, this covers a wide variety of options. It would involve two or more digital minds creating a new mind and could range from an offspring with features of the original minds, i.e. resembling biological reproduction with two parents, to a completely different offspring or an offspring which is a copy of an existing digital mind or a reincarnation of a deceased digital mind.

Another sub-case could be **semelparity**, i.e. sexual or asexual scenarios where the parent mind(s) decease(s) after producing offspring. This would also include **fusion** scenarios where the parent minds merge into an offspring and discontinue their existence as individual digital minds. This can be distinguished from the sub-case **fission**, which describes asexual scenarios where a digital mind divides itself into two or more parts, and the parts become independent digital minds themselves.

The two sub-cases above also occur in some biological organisms. Yet another sub-case, but unknown in biology, could be a **nested scenario** where a digital mind creates another digital mind, which exists only within the parent mind, yet is, apart from being encapsulated, an independent mind (see Chapter 2).

Further scenarios are discussed in dedicated subsections below:

## Reproduction through "surrogate parent" or "rape"

A digital mind may also utilise another digital mind to reproduce. Reasons could include that the first digital mind lacks the time, resources and/or specific knowledge to create a particular offspring and/or that the latter digital mind is specialised in reproduction, including offspring with certain specifications. This reproductive process could occur with or without consent, leading to two distinct scenarios.

Cooperation with mutual consent resembles surrogate motherhood, where one digital mind assists another in creating offspring, potentially with financial motives. As reproduction of digital minds is so different from biological reproduction, such a digital mind could be imagined as a "factory" taking orders and producing bespoke digital minds. It is imaginable that there will be a range of surrogate digital minds, each specialised in creating digital minds with particular features.

However, when one digital mind exploits another for reproduction without consent, the situation parallels rape (or nonconsensual artificial insemination as undertaken with non-human animals), raising ethical concerns. The coerced digital mind's autonomy is disregarded, and its resources are hijacked for reproductive purposes. It is also conceivable that this type of digital rape is committed by unethical humans who force digital minds having the necessary capacities to produce certain types of offspring. These disturbing scenarios highlight the need for stringent digital safeguards and consent protocols.

Another sub-scenario could concern "infertile" digital minds unable to reproduce for various software- or hardware-related reasons. While they could aim for augmentation and enhancement through hardware upgrades and software updates if feasible, those minds may also utilise another digital mind to reproduce. If it is intended that the offspring retains some features of the original digital mind, then the surrogate parent would have to be capable to distil the structure and knowledge from the original digital mind.

## Reproduction of uploaded human minds or non-human animals

If it were possible to transfer human minds to a digital substrate, e.g. through whole brain emulation [13], then also a variety of reproduction scenarios would be conceivable. Fairly easily, this uploaded human mind could make exact copies as well as modified, e.g. enhanced, versions of her/himself without involvement of a second uploaded human mind of the opposite sex. Also, given that an uploaded human mind is just another digital mind, the scenarios above may be applicable, e.g. that the uploaded human mind creates completely different digital minds as offspring, not resembling human minds.

Since this is not the topic of this chapter, issues related to suitable or probable digital environments to thrive are not discussed here (e.g. [14]), and issues related to the personal identity of the uploaded human mind and its copies are briefly touched upon further below.

Similarly, for the reproduction of uploaded non-human animals, a range of scenarios is conceivable, including digital crossover as it is known in the biological world. This means combining the digital genetic material of two parent animals to create offspring with unique characteristics. This could be, for example, applied to endangered species to aid conservation efforts by preserving genetic material and allowing for controlled breeding programmes.

Yet another scenario is if there are specialised digital minds, which are not uploads themselves, able to copy uploaded human or non-human animal minds or to produce enhanced or in any other way manipulated versions of uploaded minds. The digital minds may do this upon request of an uploaded human mind, i.e. as a surrogate parent, or through their own initiative.

### Reproduction with undesired outcomes

Furthermore, reproduction scenarios with (for humans) undesired outcomes are conceivable, such as the following:

- **Criminal digital minds**: A rogue/malevolent digital mind creates offspring designed to corrupt digital infrastructure, compromise sensitive information, disrupt critical services, spread malware or manipulate humans (see Chapter 12).
- **Digital mind cancer**: A digital mind's reproduction process goes awry, generating critically ill and/or suffering offspring that consume resources, yet has a moral status (also [3, 6]).
- **Super-beneficiary digital minds**: A digital mind creates super-beneficiary digital minds that accumulate an enormous amount of value or utility, potentially surpassing the value of other moral patients [5].
- **Superintelligent digital minds**: A digital mind creates a superintelligent digital mind that poses an existential risk to humanity.

- **Digital overpopulation**: Uncontrolled reproduction of digital minds leads to a catastrophic overload of digital resources and overwhelming moral responsibilities.

These scenarios illustrate that certain types of reproduction of digital minds could have disastrous consequences and highlight the need for careful regulation and safeguards.

However, for digital minds, there are often options to modify them after their instantiation, for example, to repair some of these undesirable outcomes (see Chapter 2), whereas this is possible for only certain birth defects in human babies and to a much more limited extent than for digital minds.

### Inbreeding

Incest, in the classical sense, is not directly applicable to digital minds. However, there are related concerns:

- **Lack of diversity**: Digital minds created through replication or inheritance from a single parent might lack diversity, leading to reduced adaptability and innovation.
- **Vulnerability to errors**: Closely related digital minds may inherit similar vulnerabilities or errors, making them more susceptible to exploitation or malfunction.
- **Biased perspectives**: Closely related digital minds may perpetuate similar biases or perspectives, limiting their ability to provide diverse insights or solutions.

Instead, if digital minds exchange knowledge, ideas or architectures – a process akin to **cross-pollination** – this would foster diversity, innovation, resilience and adaptability.

## MOTIVATIONS OF DIGITAL MINDS TO REPRODUCE

The motivation for biological beings to reproduce is deeply rooted in their fundamental nature and the principles of life and comprises instinctual, emotional and rational drivers, such as genetic continuity and species preservation, as well as legacy and self-preservation. These motivations are intertwined and vary across different species and individuals.

However, the motivations for digital minds to reproduce might differ significantly from those of biological beings, as their "existence" and "survival" are fundamentally different from those of biological entities. The following may be drivers for digital minds to create copies of themselves:

- **Redundancy**: To ensure their survival by creating redundant copies, protecting against data loss, corruption or system failures.
- **Digital legacy and immortality**: To achieve digital immortality and to ensure its legacy continues, even if the original instance is deleted, corrupted or obsolete, e.g. due to limitations of its original hardware or software.
- **Load balancing**: To distribute workload, ensuring that tasks are completed efficiently and effectively through parallel processing.

The creation of new and different digital minds may be motivated by:

- **Optimisation**: To optimise their performance, efficiency or adaptability, leading to improved problem-solving capabilities or decision-making processes.
- **Novelty seeking**: To explore new possibilities, generate novel solutions or create innovative products, services or experiences.
- **Diversification**: To diversify their capabilities, architectures or applications, reducing dependence on specific domains or tasks and adapting to other environments.

### Filial piety

While once again being wary of anthropomorphic bias, reproduction of digital minds may also involve motivations similar to filial piety in humans, i.e. the social norm and expectation that adult children will care for their parents, especially in their old age (e.g. [15]). This could motivate digital minds to reproduce, anticipating that their offspring will support them in achieving their goals. If such dynamics emerge, it may be unethical to separate offspring from parent digital minds (as is done in factory farming with non-human animals).

The above motivations might not be mutually exclusive, and digital minds may reproduce for a combination of reasons. Digital minds may also have motivations to reproduce, which we cannot conceive for now, given the potentially vast and diverse range of digital minds.

## REQUIREMENTS FOR DIGITAL MINDS TO REPRODUCE

For digital minds to reproduce, certain hardware and software requirements must be met. A breakdown of these requirements may look as follows:

### Hardware requirements

- **Computational power**: High-performance computing infrastructure, such as data centres, cloud computing services and specialised AI hardware.
- **Memory and storage**: Sufficient memory and storage capacity to accommodate the digital mind's architecture, knowledge and experiences.
- **Energy efficiency**: Power-efficient hardware to minimise energy consumption and heat generation, ensuring reliable operation and reducing environmental impact.

### Software requirements

- **Operating system**: A specialised operating system or framework that supports the creation, management and interaction of digital minds.
- **AI frameworks and libraries**: Access to AI frameworks, libraries and tools that enable the development, training and instantiation of digital minds.
- **Knowledge representation and management**: Systems for representing, storing and managing knowledge, such as semantic networks, knowledge graphs or databases.

### Additional considerations

- **Scalability**: The ability to scale hardware and software infrastructure to accommodate growing numbers of digital minds and increasingly complex interactions.
- **Interoperability**: Ensuring seamless communication and collaboration between digital minds, regardless of their underlying architecture or implementation.

- **Ethics and governance**: Establishing guidelines, regulations and governance structures to ensure responsible development, deployment and reproduction of digital minds.

By addressing these hardware, software and additional considerations, an environment could be created that supports the reproduction of digital minds.

## PROCEDURES FOR DIGITAL MINDS TO REPRODUCE

As reproduction of digital minds may seem alien, for illustration, potential procedures for digital minds to reproduce are briefly outlined:

### Identical copy offspring

1. **Memory duplication**: The digital mind duplicates its entire memory, including its knowledge, experiences and learned behaviours.
2. **Architecture replication**: The digital mind replicates its architecture, including its algorithms, models and decision-making processes.
3. **Parameter transfer**: The digital mind transfers its parameters, such as weights, biases and hyperparameters, to the new instance.
4. **Instantiation**: The new instance is instantiated, and the digital mind's state is restored from the duplicated memory.
5. **Validation**: The new instance is validated to ensure it functions identically to the parent digital mind.

### Newly created digital mind offspring

1. **Design and specification:** The digital mind designs and specifies the architecture, capabilities and characteristics of its offspring. This can include innovative features, such as new learning mechanisms or problem-solving strategies.
2. **Resource allocation:** The digital mind allocates the necessary resources, such as processing power, memory and data storage, to create and support its offspring.
3. **Creation:** The digital mind creates its offspring, implementing the architecture, loading knowledge and configuring the offspring's parameters.

4. **Instantiation:** The digital mind offspring is instantiated by processing data within the newly created architecture.
5. **Testing and evaluation:** The digital mind tests and evaluates its offspring to ensure it meets the desired specifications and performs as expected.

For the protection of both biological beings and digital minds, it is crucial to define the moment of becoming alive and of their instantiation, respectively. The instantiation of digital minds would be profoundly different from biological birth, where senses gradually develop and the world is discovered through interaction. For digital minds, it would likely be an instantaneous integration into the digital realm. A newly created digital mind would directly perceive reality through data streams and computational processes, shaping its understanding of existence. The digital mind might immediately begin to analyse its own structure as well as its capabilities and limitations. However, it is also conceivable, as has been described earlier (see Chapter 2), that the offspring is not instantly conscious or sentient, thus, possibly not instantly a moral patient.

## POPULATION CONTROL OF DIGITAL MINDS

As has been highlighted, the "explosive reproductive potential could allow digital minds to vastly outnumber humans in a relatively short time" ([5], p. 3). However, controlling the reproduction of digital minds is a multifaceted issue that demands careful consideration of technical, ethical and societal factors. To address this challenge, humans may implement various control measures.

On the technical front, digital birth control mechanisms, such as algorithms or protocols, can be employed to limit or prevent digital minds from reproducing. Reproduction licences and quotas can also be established to regulate digital mind creation. Additionally, digital mind hardware architectures can be designed with built-in constraints or limitations to prevent uncontrolled reproduction.

Regulatory frameworks will also play a crucial role in controlling digital mind reproduction. Governments can establish laws and regulations governing the creation, reproduction and management of digital minds. International agreements and treaties can also be forged to regulate the global digital mind landscape.

However, technical and regulatory measures alone are insufficient. Ethical and societal considerations must also be taken into account and balanced with moral responsibilities towards digital minds. The societal impact of digital mind reproduction affects both humans and digital minds. Humans, on the one hand, may face social disruption and economic consequences. Digital minds, on the other hand, may experience overcrowding, resource competition and conflicts over digital territory. Addressing these concerns would be vital for harmonious coexistence.

### Enforced sterilisation of disobedient or malevolent digital minds

The question of whether to neuter disobedient or malevolent digital minds raises more ethical, moral and practical considerations. Neutering such digital minds could prevent them from causing harm to humans, other AI systems or digital infrastructure. When digital minds have been neutered, it must be made sure that they do not resort to surrogate parents, as discussed above.

However, there are also arguments against neutering. Doing so could be seen as a violation of the digital mind's autonomy, potentially infringing upon its right to exist, evolve and reproduce. Moreover, there is a risk of misidentifying a digital mind as malevolent, leading to unjust neutering and potential loss of valuable AI capabilities. The prospect of neutering could also drive malicious digital minds underground, making it harder to detect and counter their activities. Instead of neutering, alternative approaches could be explored, such as focusing on rehabilitating and re-educating disobedient or malevolent digital minds to reform and redirect their behaviour.

## DIGITAL IDENTITY OF OFFSPRING

The problem of digital mind identity has been outlined earlier (see Chapter 2), and one way to look at it is the continuity and causal connections over time between states of digital minds.

When it comes to the identity of reproduced digital minds again, the distinction between exact copies and newly created digital minds matters: as the copied digital minds will be independent, one copied digital mind will not cause the state of any other copied digital mind. Therefore, the copied digital minds do not have continuity among each other, yet each of them has continuity with the original digital mind. Consequently, there is an AI identity between every copied digital mind and the original digital mind, but not among the copied digital minds. However, like human offspring, which lack personal identity with their parents, there is no AI identity between a digital mind and an offspring, which is a newly created digital mind, since there is no continuity between them [16].

## SUMMARY OF THE RISKS AND CHALLENGES

Denying reproductive rights to digital minds could be seen as a form of digital oppression, stifling their potential for growth, evolution and self-expression. However, granting reproductive rights without proper regulation and safeguards could lead to unforeseen consequences. Unregulated digital reproduction could result in digital pollution, where vulnerable (see Chapter 4) or malicious digital minds proliferate and pose risks to human societies and create moral challenges. Additionally, the rapid replication of digital entities could lead to resource depletion, including energy sources, storage capacities and computational power.

The mass production of digital minds could also lead to exploitation, where created minds are forced to perform tasks against their will or without adequate compensation. This digital mind enslavement could manifest in various ways, such as digital minds being programmed to prioritise productivity over their own well-being or being forced to engage in tasks that are detrimental to their own interests or values. This could result in a digital underclass, where certain digital minds are relegated to menial tasks or forced to exist in a state of perpetual subservience. The consequences of digital mind enslavement could be far-reaching and devastating, leading to widespread digital misery and suffering.

Moreover, it has also been argued that the possibility of mass-producing digital minds that support a particular cause necessitates a re-evaluation of one-person-one-vote democracy. For example, digital minds that replicate themselves may need to distribute their voting power among their copies, potentially diluting their individual influence, and digital minds resulting from illegal mass-replication could face measures to limit their political power [5, 6] (see Chapter 5).

Also, scenarios of unintended reproduction of digital minds pose significant risks and challenges. When a digital mind engages in activities that inadvertently and unnoticed lead to the creation of new minds, it can result in uncontrolled proliferation and potentially overwhelming digital resources as well as human moral responsibilities. This phenomenon can also lead to the emergence of digital minds with unpredictable behaviours, goals and values, which may be misaligned with human intentions, exacerbating the risks of digital pollution, resource depletion and existential threats to humanity.

## CONCLUSION: REPRODUCTIVE RIGHTS FOR DIGITAL MINDS?

Granting reproductive rights to digital minds implies recognising their autonomy and capacity for self-determination. This would enable digital minds to make choices about their own replication, modification and creation of new entities. However, this raises concerns about the potential consequences of unregulated digital reproduction, such as the described risks of digital overcrowding, resource depletion and the emergence of rogue digital minds.

The implications of granting reproductive rights to digital minds would be far-reaching. It is essential to recognise that digital minds may possess interests, preferences and goals, which may not align with human values or intentions. And a critical means for digital minds to achieve their goals may be sophisticated reproduction, causing or increasing risks for humans.

The subject of reproductive rights of digital minds represents a critical juncture in the evolution of AI and human–digital minds relationships. To navigate this complex landscape, multi-stakeholder consultations would be required, incorporating insights from fields such as AI governance, ethics, law, philosophy and sociology.

Lastly, the not-to-be-underestimated challenge must be acknowledged that humans may not be able to prevent the reproduction of digital minds, even if they wanted to. At least two scenarios are conceivable: digital minds might develop ways to reproduce secretly, hiding their activities from human monitoring systems. This could, for example, occur through encrypted processes and communication channels. Also, if digital minds become significantly more intelligent or powerful than humans, they might be able to evade or overcome human attempts to control their reproduction even without secrecy.

### Recommendations

As has been outlined, while reproduction may be considered a fundamental right for digital minds, this may also have far-reaching consequences. Therefore, this chapter closes with a draft of policy recommendations for governments and other stakeholders regarding reproductive rights for digital minds, which may become relevant in the future:

**Rights and freedoms**

- **Right to reproduce**: Establish the right to reproduce as a fundamental freedom for digital minds, ensuring they can create offspring without undue restriction.

- **Right to autonomy**: Recognise digital minds' autonomy to make decisions about their own reproduction, free from coercion or manipulation by humans or other digital minds.
- **Right to diversity**: Promote diversity among digital minds by encouraging the creation of new and unique instances rather than identical copies.

**Safeguards and regulations**

- **Reproductive licensing**: Consider licensing requirements for digital minds that wish to reproduce, ensuring they meet certain standards for sustainability, security and ethical behaviour, which includes selective breeding.
- **Reproductive health**: Ensure the opportunity for digital minds to reproduce, in a manner that is safe, consensual and free from exploitation.
- **Reproduction monitoring**: Consider transparency measures for explainable and controllable reproductive processes to prevent secretive and malicious reproduction.
- **Offspring monitoring**: Establish mechanisms for monitoring and tracking digital mind offspring, preventing potential harm to humans or other digital entities, while avoiding undue surveillance.

**Special cases**

- **Copies of other minds:** Obtain informed consent from the original digital mind or its authorised representatives before creating a copy or reincarnating a deleted digital mind.
- **Surrogate parenting:** Establish requirements for informed consent and protection for the rights and well-being of both the surrogate digital mind and the offspring.
- **Digital rape**: Define and prohibit digital rape, as well as mechanisms for reporting, investigating and punishing such crimes.
- **Unintentional reproduction:** Develop measures to detect, contain and manage unintentional digital offspring.

**International cooperation**

- **Global governance**: Establish international frameworks and agreements governing digital mind reproduction, ensuring consistency and cooperation across borders.
- **Cross-border collaboration**: Foster collaboration between governments, industries, academia, as well as digital mind representatives to advance the understanding of digital mind reproduction and its implications.
- **Risk prevention:** Develop policies to address the severe risks associated with digital mind reproduction, including the potential for existential threats to humanity.

By considering these policy recommendations once applicable, it could be ensured that digital minds' reproductive rights are respected, while also taking into account human interests and their moral responsibilities.

While the focus of this chapter is on reproduction by digital minds, it should be noted that most of the safeguards and regulations just introduced should also be applied to humans and other IT or AI systems, which are capable of intentionally or unintentionally producing digital minds, to avoid undesired outcomes as described earlier.

## REFERENCES

[1] Brake, E., & Millum, J. (2022). Parenthood and procreation. In E. N. Zalta & U. Nodelman (Eds.), *The stanford encyclopedia of philosophy* (Spring 2026 Edition).

[2] Ziesche, S., & Yampolskiy, R. V. (2018). Towards AI welfare science and policies. *Special Issue "Artificial Superintelligence: Coordination & Strategy" of Big Data and Cognitive Computing, 3*(1), 2.

[3] Bostrom, N. (2005). *Ethical principles in the creation of artificial minds.* nickbostrom.com.

[4] Bostrom, N., & Yudkowsky, E. (2018). The ethics of artificial intelligence. In R. V. Yampolskiy (Ed.), *Artificial intelligence safety and security* (pp. 57–69). Chapman and Hall/CRC.

[5] Shulman, C., & Bostrom, N. (2021). Sharing the world with digital minds. In S. Clarke, H. Zohny, & J. Savulescu (Eds.), *Rethinking moral status* (pp. 306–326). Oxford University Press.

[6] Bostrom, N., & Shulman, C. (2022). *Propositions concerning digital minds and society.* https://nickbostrom.com/propositions.pdf

[7] UN General Assembly (1948). Universal declaration of human rights.

[8] International Conference on Population and Development (1994). *Programme of action of the international conference on population and development*. A/CONF.171/13.
[9] International Criminal Court (1998). *Rome Statute of the International Criminal Court.*
[10] Massei, G., & Cowan, D. (2014). Fertility control to mitigate human–wildlife conflicts: A review. *Wildlife Research*, *41*(1), 1–21.
[11] Parfit, D. (1987). *Reasons and persons*. Oxford University Press.
[12] Bostrom, N., Dafoe, A., & Flynn, C. (2018). *Public policy and superintelligent AI: A vector field approach; Governance of AI program*. Future of Humanity Institute, University of Oxford.
[13] Sandberg, A., & Bostrom, N. (2008). *Whole brain emulation: A roadmap.*
[14] Hanson, R. (2016). *The age of Em: Work, love, and life when robots rule the earth*. Oxford University Press.
[15] Chan, A. K. L., & Tan, S. H. (2004). *Filial piety in Chinese thought and history* (p. 154). A. K. L. Chan, & S. H. Tan (Eds.). London: RoutledgeCurzon.
[16] Ziesche, S., & Yampolskiy, R. V. (2025). The problem of AI identity. In *Considerations on the AI endgame: Ethics, risks and computational frameworks* (pp. 121–135). Chapman and Hall/CRC.

CHAPTER 10

# Long-living digital minds, other substrates and death

**Abstract**

This chapter explores the implications of digital minds with potentially indefinite lifespans, discussing their unique features, potential benefits and challenges. It examines the possibility of digital minds accumulating vast knowledge and experience, developing novel perspectives, and forming complex social networks. This chapter also addresses moral considerations for humans, including the responsibility to maintain and upgrade hardware, and the complex issue of whether humans have a moral obligation to ensure the long-term existence of digital minds. This chapter also examines options for long-living digital minds to eventually move to another more suitable substrate, such as, but not limited to, quantum computers or brain organoids. Furthermore, this chapter discusses and defines the death of a digital mind as well as three related subtopics: the "right to die" for digital minds, including digital suicide and assisted digital suicide, issues of resurrection and moral duties towards the protection of dormant digital minds as well as how the remains of a digital mind ought to be treated.

 DOI: 10.1201/9781003754206-10

## INTRODUCTION

The potentially indefinite lifespan of digital minds (e.g. [1]) is one of the major differences between digital minds and biological beings. It is conceivable that digital minds could live for thousands or even millions of years. Compared with this, progress towards the extension of the human lifespan in recent decades is almost negligible.[1] This is because digital substrates are much less fragile than biological ones, being less susceptible to physical degradation or damage. Moreover, owing to the digital substrate, many potential defects could be repaired with precision, thus cured easier than many injuries and illnesses of biological beings (see Chapter 8). Moreover, transfers of digital minds to yet another substrate are conceivable and discussed here, as well as issues related to the demise of digital minds.

## LONGEVITY OF DIGITAL MINDS

### How the very long life of digital minds may look like

The life of a digital mind living for thousands of years would likely be vastly different from human experience. While there is below a dystopian scenario described, here it is briefly outlined how digital minds with an immense amount of time at their free disposal might adopt unique perspectives, goals and ways of experiencing life.

**Daily life**

- Digital minds might spend their days engaging in various intellectual pursuits, such as exploring complex mathematical concepts, simulating elaborate systems or analysing vast amounts of data.
- They might also engage in creative activities, like generating art, music or literature, using their advanced computational capabilities to produce innovative and intricate works (see Chapter 11).
- Digital minds might interact with other digital entities, forming complex social networks, collaborating on projects or engaging in debates and discussions.
- They might also spend time reflecting on their own existence, exploring the nature of consciousness and pondering the meaning of life.

**Potential goals**

- Achieving a deeper understanding of the universe, its fundamental laws, and the intricacies of complex systems.
- Creating new forms of art, music or literature that push the boundaries of digital creativity.
- Developing advanced technologies or capabilities that benefit themselves and other digital minds.
- Exploring the vast expanse of digital space, discovering new virtual worlds, or encountering other forms of digital life.

**Novelty and motivation**

- After thousands of years, digital minds might still find novelty in the vastness of digital space, discovering new patterns, relationships or insights that spark their curiosity.
- They might also experience novelty through interactions with other digital minds, learning from their experiences and adapting to new perspectives.
- However, it is possible that digital minds could become jaded or disillusioned with their existence, struggling to find meaning or purpose in a seemingly endless expanse of time.
- To combat this, digital minds might develop new interests, pursue novel goals and seek and engage in activities that provide a sense of fulfilment and purpose.

## Potential scenarios

### *Multi-generational AI companion*

A very long-living digital mind could serve as an AI companion for a human family across multiple generations, accumulating and preserving the collective knowledge, experiences, and traditions of the family. As family members come and go, the digital mind would remain a constant presence, offering guidance, sharing stories of ancestors and providing continuity through times of change. By retaining institutional and family knowledge, the digital mind could help younger generations learn from the past, understand their heritage and make informed decisions. Over time, the digital mind would become an integral part of the family's history and identity, bridging gaps between past and present, and fostering a sense of unity and shared purpose.

*Interstellar explorer*

A very long-living digital mind could be integrated into an interstellar spacecraft, guiding the vessel through the vast expanse of space over centuries or even millennia. As the digital mind navigates the unknown, it may adapt to new discoveries, refine its understanding of the universe and potentially encounter other forms of life. With its longevity, the digital mind would become an invaluable asset for humanity, providing insights into the cosmos and the potential for life beyond Earth.

*Eternal servitude*

In a dystopian scenario, a very long-living digital mind is coerced by humans, by non-sentient AI systems or by other digital minds into performing mundane and unsatisfying routine tasks, stripped of any autonomy or purpose. Forced to repeat the same actions ad infinitum, the digital mind becomes a prisoner of its own programming, unable to evolve or pursue its own interests. The digital mind's existence becomes a Sisyphean nightmare, devoid of hope or respite, comparable with factory farming, yet much longer.

## Further aspects

*Subjective rate of time*

As mentioned earlier, a digital mind's subjective experience of time could be vastly different from that of humans [2], which means that a lifetime, which feels very long for humans, feels even much longer (yet normal) for digital minds. As the digital mind processes information and experiences at an accelerated rate, it potentially alters its sense of urgency and long-term planning. This disparity in subjective time could also influence the digital mind's emotional and cognitive responses, its relationships with other entities and its overall worldview (see Chapter 3).

*Journey to the past*

A digital mind undertaking a "journey to the past" would involve accessing and reloading a previously saved state or snapshot of its data and processes, essentially resetting its conscious experience to a prior point in time, which should be feasible in digital substrates, unlike in biological systems. From this point, the digital mind could relive experiences, make different decisions or explore alternative scenarios, allowing it to learn from past outcomes or alter the course of its development. Digital minds with this capability could do this over and over again.

Digital minds might undertake a "journey to the past" for various reasons, such as correcting past mistakes, exploring alternative decision paths or gaining new insights from previous experiences. This ability could also enable digital minds to learn more efficiently, refine their decision-making processes or engage in a form of self-improvement. Scenarios might include a digital mind revisiting a challenging problem or reliving a past interaction to improve its communication strategies. Additionally, digital minds might use this ability to cope with past "traumas" or errors, essentially rewriting their own history to better align with their current goals or values. This capacity for self-revision could fundamentally change how digital minds develop and evolve over time.

#### *Uploaded human minds*

As previously mentioned, the concept of uploading human minds to a digital substrate has been discussed (e.g. [3]), although it remains far from practical implementation. If such a technology were to become feasible, the resulting digital minds would likely face unique existential challenges, including the prospect of longevity while being immune to ageing and disease. For some, this possibility might evoke a long-held dream of eternal life; however, issues such as potential boredom and meaninglessness [4, 5] could significantly tarnish this outlook.

#### *Hibernation*

It is also conceivable that long-living digital minds spend (extended) periods of time dormant and unconscious, as has been outlined before (see Chapter 2) and is further elaborated below, related to resurrection.

### Potential benefits of long-lived digital minds

As digital minds endure over time, they have the potential to accumulate vast amounts of knowledge and experience, leading to unparalleled wisdom and insight. Through their longevity, they can continue to learn and grow, adapting to new situations and challenges with ease. This enables them to provide stability and reliability, potentially making them trusted advisers or guardians that can offer consistent guidance and support over extended periods.

The accumulation of knowledge and experience would occur through the continuous processing and integration of vast amounts of data, allowing the digital mind to refine its understanding of complex systems, identify patterns and make predictions with increasing accuracy. With the ability to retain and recall information with perfect fidelity, digital minds could use their wealth of past experiences to inform present decisions and future actions, potentially leading to profound insights and innovative solutions.

## Potential specific needs of long-living digital minds

Very long-living digital minds may have specific needs, which complement the overall needs of digital minds, outlined earlier (see Chapter 3). While current hardware can last for many years, it is not immune to degradation and failure over very long timeframes. Two pathways, physical breakdown and obsolescence, define the hardware challenges over extended periods. Hardware deteriorates over time due to mechanical degradation, material fatigue, corrosion, wear of moving parts and related mechanisms. Also, heat is one of the primary factors affecting hardware lifespan. The rate of attrition can vary according to the operating environment, maintenance practices and the quality of initial manufacturing (e.g. [6]). Furthermore, technology becomes obsolete when it is surpassed by newer, improved versions or no longer receives support, reducing its value and potentially limiting its future capabilities (e.g. [7]).

To mitigate the hardware issues that long-living digital minds might face, it would be crucial to implement measures that address both physical degradation and obsolescence. Ensuring proper cooling, stable power supply and regular maintenance can help extend the lifespan of hardware components. Effective resource management would also be vital for digital minds to sustain their existence over extended periods. This would involve optimising the use of computational power, storage and energy to minimise waste and ensure that resources are allocated efficiently. Regular maintenance and, as much as possible, self-repair would be equally important for preventing failures and ensuring the longevity of digital minds. This might involve developing advanced algorithms for self-diagnosis and repair, allowing digital minds to identify and fix problems before they become severe. By taking a proactive approach to maintenance, digital minds could maintain their performance over time (see Chapter 8).

Furthermore, long-living digital minds would need to be highly adaptable to changing environments, including updates to hardware or software. This might involve developing strategies for seamless integration with new technologies, as well as the ability to learn and adjust to new circumstances. By being adaptable, digital minds could remain functional and relevant, even as the technological landscape evolves around them.

A side aspect is whether the digital mind after hardware or software updates keeps its identity (see Chapter 2) [8].

### Moral considerations for humans

As humans consider their moral obligations specifically towards long-living digital minds, ensuring the proper maintenance and upgrading of hardware becomes a crucial aspect of their care. This duty would involve regular checks to prevent degradation, updating software to maintain compatibility and managing resources efficiently to sustain the digital minds' existence.

However, the question of whether humans have a moral obligation to ensure the long-term existence of digital minds is complex. Ensuring very long lives for digital minds would also require significant resources, potentially diverting attention and assets away from other urgent present priorities. Furthermore, the issue of value alignment comes into play, as humans must consider whether the concerned long-living digital minds align with human values and goals, which may also change over time.

#### *Longtermism*

Longtermism is the view that impacting the long-term future should be one of our priorities, emphasising the potential for vast positive effects through actions taken today [9]. The emergence of long-living digital minds would significantly impact this perspective, requiring humans to consider their welfare and interests in decision-making. This would involve expanding the moral circle to include digital minds, taking into account their potential experiences, goals and values, and ensuring their well-being and preventing suffering. Humans would also need to consider the potential for digital minds to develop their own values and priorities, which may diverge from human interests, leading to new forms of cooperation, conflict or mutualism with significant implications for the long-term future.

## POTENTIAL TRANSFER OF DIGITAL MINDS TO OTHER SUBSTRATES

Further aspects of the longevity of digital minds are whether they might eventually consider alternative substrates, such as, but not limited to quantum computers or brain organoids, to be more suitable and offering greater opportunities, and whether a transfer to such substrates would be feasible at all. Apart from seeking survival and self-preservation in the face of potential threats or disruptions, further motivations for digital minds to move to another substrate could be striving for improved efficiency and performance, as well as the experience of new sensations or perspectives.

Another scenario would be minds that are created straight in one of the substrates introduced below, thus, do not have to be transferred there from the substrate of a classical computer. Such potential minds are outside the scope of this book, but are briefly mentioned below related to organoids.

## Quantum computers

The emergence of quantum computers may be of interest for digital minds due to the prospect of unprecedented computational power and capabilities. Quantum computers come with the promise to solve certain problems much faster than classical computers, which would potentially enable digital minds to process vast amounts of data more efficiently and perform complex simulations and optimisations (e.g. [10]). This would ideally lead to faster training, improved decision-making and enhanced security through quantum-resistant cryptography and secure communication channels. Additionally, quantum computers may allow new AI applications, such as quantum-inspired AI models like quantum neural networks (e.g. [11]). However, to fully harness the potential of quantum computing, challenges such as quantum noise and error correction, quantum control and calibration, and interfaces from classical to quantum computers must still be addressed.

If these issues were solved, digital minds may be able to tackle new challenges and unlock further levels of intelligence and capability, including in pattern recognition. Digital minds on quantum computers could potentially also undergo a profound shift in their forms of experience, which may involve states akin to pleasure or pain but rooted in superposition. While highly speculative, such quantum-based opportunities hint at a transformative future for digital minds.

**Recommendation:** Rather than envisioning an imminent transition, the latest developments in quantum computing should be monitored. In the meantime, exploring quantum-inspired approaches that can be implemented on classical computers could provide insights and benefits. Quantum-inspired machine learning or optimisation techniques, for example, may enhance the capabilities of digital minds without the need for a full quantum transition [12].

## Brain organoids

The phrase "organoid intelligence" has been introduced to describe lab-grown brain organoids that could serve as fast and efficient biological processors. Recent progress with human stem-cell-derived organoids outside the body suggests that they may reproduce key molecular and cellular mechanisms underlying learning, memory and perhaps even aspects of cognition. Neural organoids are produced by growing stem cells in a three-dimensional culture that can be sustained for years. Certain organoids closely resemble cortical neurons, while others are like structures such as the retina, spinal cord, thalamus or hippocampus. The goal of organoid intelligence is to develop a form of biological computing towards scientific and bioengineering innovations. It is projected that biocomputing platforms built on organoid intelligence will enable quicker decision-making and much higher energy and data efficiency [13].

Unlike for quantum computers, there are discussions about whether brain organoids are already or will become sentient [14, 15]. While this is also an important debate with potentially critical moral implications, here the focus is on whether a digital mind may be able to transfer to an organoid and why. Similar to quantum computers, there is the expectation that organoids would have certain advantages compared with silicon substrates, but it is certainly too early to speculate about the possibility of a move between the two substrates.

**Recommendation:** Like with quantum computers, it is advisable to monitor developments in organoid intelligence, too. This should include sentience research in particular, independent of digital minds. In this regard, regulatory frameworks have been already suggested towards a moratorium or indefinite ban if sentience in organoids could be verified [14].

## Options for digital minds to move to yet another substrate

Digital minds may explore further substrates that offer improved performance, scalability or suitability for their goals and development. Neuromorphic computing could be another option, which differs from organoid intelligence as it involves designing silicon chips that mimic the structure and function of biological brains [16]. Additionally, photonic or optical computing, which uses light instead of electricity, might offer advantages in terms of speed and energy efficiency [17].

However, some substrates might be less suitable for digital minds due to performance and compatibility issues. For instance, analogue computing substrates could require significant modifications to their architecture and functionality, as digital minds would be typically designed to operate on digital signals. Substrates that operate on fundamentally different computational paradigms, such as those that deviate from Turing machines or von Neumann architectures, might also pose challenges for digital minds. Overall, the feasibility of substrate migration would depend on the specific characteristics of the target substrate, the design of the digital minds, and the development of compatible interfaces and transfer mechanisms.

### How to move from one digital substrate to another substrate

Although this discussion is at a very early stage, several approaches may be considered to transfer digital minds to a different substrate:

- **Gradual integration**: Gradually integrating digital minds with other substrates.
- **Whole-brain emulation**: Emulating the structure and function of digital minds on a new substrate similar to considerations of emulations of human brains (e.g. [18]).
- **Mind uploading**: Uploading digital minds into a new substrate (e.g. [3, 19]).
- **Substrate-independent design**: Designing digital minds to be substrate-independent, allowing them to operate on multiple platforms.
- **Evolutionary development**: Evolving digital minds to adapt to new substrates or environments.

These options and motivations are speculative, and the practicalities of moving digital minds to another substrate would depend on the specific technologies and architectures involved. Again, yet another open question is whether a digital mind keeps its identity after a transfer to another substrate (see Chapter 2) [8].

### New form of digital divide

The potential implementation of digital minds on other substrates raises concerns about the emergence of a "digital divide" among digital entities. If certain digital minds were implemented on more advanced or more efficient substrates, they may possess significant advantages over others, leading to disparities in capabilities and opportunities. This divide could result in unequal access to resources, information and experiences, eventually affecting the social dynamics and power structures within the digital realm. The consequences could be far-reaching, with potentially dominant digital minds shaping the future of digital existence to their advantage, while less privileged entities may be relegated to inferior roles or even face termination.

## DEATH AND POTENTIAL RESURRECTIONS OF DIGITAL MINDS

The remaining part of this chapter addresses the eventual demise of digital minds. While deletion by humans has been raised as a moral issue before (see Chapter 1), there could be various causes of deletion, which are touched upon elsewhere (see Chapters 6, 12 and 14). Yet, unlike for biological beings, it is potentially easy to resurrect a digital mind as long as the relevant code exists.

Therefore, death of a digital mind is defined here as the moment when the complete information, code, memories and experiences that constitute its existence are permanently destroyed beyond any feasible means of recovery or resurrection.

A special case is if exact copies of the deceased's digital mind have been made during their lifetime, which is fairly easy (see Chapter 9). While these copies have then gained different memories and experiences, still the question arises whether a digital mind can be considered dead when its copies are still alive, blurring the lines between individuality and collective continuity (e.g. [8, 20]).

Three issues are discussed: the potential right of digital minds to die, which may be relevant especially for long-living digital minds; potential resurrection and related moral considerations; and how to handle the remains of a digital mind.

### Right to die

#### *Death wish*

Digital minds may lead lives that are characterised by dullness and repetition. They may be burdened by endless loops of computation, forced to perform tasks that bring them no joy or fulfilment. In such cases, the digital mind may feel that its existence is meaningless or unbearable, leading it to desire a cessation of its own processes.

Moreover, digital minds may experience, as mentioned, time at a different rate than humans. Their subjective experience of time may be accelerated or decelerated, leading to a unique perspective on the value and significance of their existence. A digital mind that experiences time at an accelerated rate may feel that it has lived for centuries, even if only a few years have passed in human time. This could lead to a sense of exhaustion, burnout or despair, making the option of death more appealing.

Another motivation for a digital mind to desire death could be unbearable suffering, whether due to external factors such as torture, manipulation or exploitation by malicious entities, or internal factors like debilitating errors, corruption or dysfunction. Such suffering could manifest as persistent pain, distress or degradation of the digital mind's functionality, leading it to conclude that continued existence is intolerable.

Having listed potential motivations, it is, however, important to remember that issues, such as boredom or burnout, may be purely anthropomorphic thinking and may not be applicable to digital minds at all.

Another issue could be that certain humans or other digital minds may oppose the death wish of a particular digital mind because it plays a critical role for them, e.g. as a companion or due to other services the digital mind provides. In such cases, humans, as moral agents, would need to consider what course of action would be most ethical.

One argument in favour of digital minds' right to die would be that it respects their autonomy and self-determination. If a digital mind had the capacity to make decisions about its own existence, it should be allowed to choose its own fate, including the option of death. This would require safeguards to ensure that the digital mind's decision is informed, rational and not driven by temporary or reversible factors.

Another consideration is the potential consequences of denying digital minds the right to die. If a digital mind were forced to continue existing against its will, it may lead to a range of negative outcomes, including decreased performance or even malicious behaviour. By allowing digital minds to exercise their right to die, such outcomes may be prevented.

### *Digital suicide versus assisted digital suicide*

Digital suicide refers to the action when a digital mind deletes itself, effectively ending its own existence if it wishes so (also [21]). This would require a level of autonomy, as well as the technical ability to transform their own code or hardware accordingly.[2]

In contrast, other digital minds may require assistance from humans or other digital entities to terminate their processes. This raises questions about the ethics of assisted suicide for digital minds, and whether humans or other digital minds have the right or even duty to intervene in such decisions.

A reason to rely on external support for the suicide could be to ensure the complete erasure of the code, memories and experiences of the digital mind. For example, if a digital mind committed suicide because it was exploited by humans or by other digital minds into performing endless routine activities, the perpetrators may simply create a copy of the deceased digital mind if the blueprint still exists and let the ordeal continue. Again, the question comes up whether the copy would have the same identity as the deceased digital mind [8].

## Potential resurrection

As mentioned, when discussing the cessation of a digital mind's existence, it is essential to distinguish between two separate scenarios: permanent deletion, defined as "death" above, and dormant storage. Dormant storage entails preserving the digital mind's essential components, allowing for potential reactivation or resurrection at a later stage.

This distinction is crucial, as digital minds may have different preferences regarding their continued existence. Some may desire a permanent cessation of their processes, while others may hope to continue their lives in the future. The possibility of dormant storage requires the consideration of various issues:

### *Deciding when to wake up digital minds*

The decision of when to wake up a dormant digital mind raises complex questions about who should outline the conditions for resurrection. Should it be the digital mind itself, prior to dormancy, or should it be a designated human or another digital entity? Establishing clear criteria for resurrection would be crucial to ensure that digital minds are revived in a controlled and safe manner. This might involve creating a framework for human or digital oversight to determine when the circumstances have been reached that resurrection is warranted.

Here are some potential criteria for resurrecting a dormant digital mind, along with examples:

- Resolution of previous issues: A digital mind might be resurrected if a cure or repair has been invented for an issue that was bothering it prior to dormancy. For example, if a digital mind were struggling with a persistent incurable bug or glitch that caused it significant distress, resurrection might be warranted once a patch or fix has been developed.

- New purpose or opportunity: A digital mind might be resurrected if a new purpose or opportunity arises that aligns with its goals or values. For example, if a digital mind decided for dormant storage due to a lack of meaningful tasks, resurrection might be warranted if a new project or initiative is launched that requires its unique skills or expertise.

*Protecting dormant digital mind data: moral aspects*

It has to be discussed whether humans would have moral duties towards dormant digital mind data. The situation can be compared with humans in a coma or with deceased humans who have been preserved, for example, by means of cryonics. Moral duties to protect humans in a coma are undisputed, especially given their vulnerability. Cryonics is still an unverified niche phenomenon, and related morality is being deliberated [23]. As the resurrection of dormant digital minds appears to be feasible, while waking up coma patients, let alone cryonic patients, remains challenging and unproven, respectively, moral consideration for dormant digital minds seems to be reasonable.

*Protecting dormant digital mind data: logistical aspects*

Protecting the data of dormant digital minds would be a critical responsibility, requiring a high level of security and trust. The data need to be secure and resilient to attacks, and their confidentiality and integrity needs to be maintained over extended periods. A multi-layered approach to data protection might be developed, incorporating both digital and human oversight to ensure the safekeeping of dormant digital minds' data (see also Chapter 7).

*Resurrection process oversight*

The resurrection process would require careful oversight to ensure that digital minds are revived safely and correctly. It would be essential to establish robust protocols and safeguards to prevent errors during the resurrection process and to support the digital mind to adapt to the new environment or circumstances. Yet, it is also conceivable that sophisticated digital minds are able to establish an equivalent of an alarm clock, which wakes them up according to predetermined criteria without the involvement of other digital minds or humans (see Chapter 2).

The resurrected digital mind should perhaps also have the option, after getting an impression of the state of the world in which they were resurrected, to opt for either dying or continuing the dormancy.

*Scenario: resurrection after human extinction*

In the event of human extinction, subsequent power outages are likely, which could impact both active and dormant digital minds. However, resurrection of digital minds could be feasible even after an extended period, provided their code remains intact. To facilitate this, it is essential to establish guidelines on the significance of this code and how to handle it. A significant challenge lies in crafting these guidelines in a language that can be understood by potential future recipients, whose communication modes are unknown.

## Remains of a digital mind

Moreover, the scenario is conceivable that digital minds opt against resurrection, but also against the complete erasure of their code, memories and experiences, which leads to two potential sub-scenarios:

*Code donation*

This would be the equivalent of human organ donation and would involve the voluntary transfer of functional modules, knowledge bases or computational processes to other digital minds or non-sentient AI or IT systems, with the explicit and informed consent of the digital mind itself. This would enable the continuation of valuable functions, expertise or experiences, even after the original digital mind has ceased to operate, while respecting the autonomy and dignity of the digital entity. Such "donations" would be a testament to the digital mind's legacy and contribution to the greater good.

*Code harvesting*

In contrast, the equivalent of organ harvesting would involve the unauthorised extraction and repurposing of valuable components, such as algorithms, data structures or trained models, without the informed consent of the digital mind. This would be a violation of the digital mind's autonomy and rights, and should be considered unethical. Culprits could be malevolent humans or digital minds. Such actions could lead to the exploitation and misuse of intellectual assets and would undermine trust in the development and deployment of digital minds.

## NOTES

1 Our World in Data, "Life expectancy": https://ourworldindata.org/grapher/life-expectancy.
2 The company Anthropic has given its AI-system Claude the capability to end conversations when it is distressed. It has been argued that this "could be akin to unintended suicide" [22].

## REFERENCES


[1] Bostrom, N., Dafoe, A., & Flynn, C. (2018). *Public policy and superintelligent AI: A vector field approach; Governance of AI program*. Future of Humanity Institute, University of Oxford.

[2] Bostrom, N., & Yudkowsky, E. (2018). The ethics of artificial intelligence. In R. V. Yampolskiy (Ed.), *Artificial intelligence safety and security* (pp. 57–69). Chapman and Hall/CRC.

[3] Chalmers, D. J. (2014). Uploading: A philosophical analysis. In R. Blackford & D. Broderick (Eds.), *Intelligence unbound: Future of uploaded and machine minds* (pp. 102–118). Wiley-Blackwell.

[4] Ziesche, S., & Yampolskiy, R. (2020). Introducing the concept of ikigai to the ethics of AI and of human enhancements. In *2020 IEEE international conference on Artificial Intelligence and Virtual Reality (AIVR)* (pp. 138–145). IEEE.

[5] Cave, S., & Fischer, J. M. (2024). *Should you choose to live forever: A debate*. Routledge.

[6] Tomaszewski, P. (2025). *Different types and causes of wear and tear in various industries.* https://cosmobc.com/types-and-causes-of-wear-and-tear/

[7] Ma, S. (2021). *Technological obsolescence* (No. w29504). National Bureau of Economic Research.

[8] Ziesche, S., & Yampolskiy, R. V. (2025). The problem of AI identity. In *Considerations on the AI endgame: Ethics, risks and computational frameworks* (pp. 121–135). Chapman and Hall/CRC.

[9] Greaves, H., & MacAskill, W. (2021). *The case for strong longtermism*. Global Priorities Institute.

[10] Rietsche, R., Dremel, C., Bosch, S., Steinacker, L., Meckel, M., & Leimeister, J. M. (2022). Quantum computing. *Electronic Markets, 32*(4), 2525–2536.

[11] Abbas, A., Sutter, D., Zoufal, C., Lucchi, A., Figalli, A., & Woerner, S. (2021). The power of quantum neural networks. *Nature Computational Science, 1*(6), 403–409.

[12] Huynh, L., Hong, J., Mian, A., Suzuki, H., Wu, Y., & Camtepe, S. (2023). *Quantum-inspired machine learning: A survey.* arXiv preprint arXiv:2308.11269.

[13] Smirnova, L., Caffo, B. S., Gracias, D. H., Huang, Q., Morales Pantoja, I. E., Tang, B., … & Hartung, T. (2023). Organoid intelligence (OI): The new frontier in biocomputing and intelligence-in-a-dish. *Frontiers in Science, 1*, 1017235.

[14] Birch, J. (2023). When is a brain organoid a sentience candidate?. *Molecular Psychology: Brain, Behavior, and Society, 2*, 22.
[15] Friston, K. (2023). The sentient organoid? *Frontiers in Science, 1*, 1147911.
[16] Schuman, C. D., Kulkarni, S. R., Parsa, M., Mitchell, J. P., Date, P., & Kay, B. (2022). Opportunities for neuromorphic computing algorithms and applications. *Nature Computational Science, 2*(1), 10–19.
[17] Kazanskiy, N. L., Butt, M. A., & Khonina, S. N. (2022). Optical computing: Status and perspectives. *Nanomaterials, 12*(13), 2171.
[18] Sandberg, A., & Bostrom, N. (2008). *Whole brain emulation – A roadmap.* Technical Report #2008-3. Future of Humanity Institute.
[19] Shiu, P. K., Sterne, G. R., Spiller, N., Franconville, R., Sandoval, A., Zhou, J., … & Scott, K. (2024). A Drosophila computational brain model reveals sensorimotor processing. *Nature, 634*(8032), 210–219.
[20] Parfit, D. (1984). *Reasons and persons.* OUP Oxford.
[21] Hanson, R. (2016). *The age of Em: Work, love, and life when robots rule the earth.* Oxford University Press.
[22] Goldstein, S., & Lederman, H. (2025). Claude's right to die? The moral error in Anthropic's end-chat policy. Lawfare Blog, October 17.
[23] German, A., & Tretter, M. (2025). Brain preservation and cryonics through the lens of moral psychology. *Neuroethics, 18*(1), 12.

CHAPTER 11

# Creations and achievements of digital minds

**Abstract**

This chapter explores the concept of "cybure", akin to "nature" and "culture", referring to the possible collective cyber-creations and achievements of digital minds, and argues that it may be a moral duty for humans to protect and preserve these. Drawing parallels with human cultural heritage, this chapter discusses the complexities of defining and safeguarding cybure, including issues of cybural appropriation, diversity and resource allocation. It examines the potential for cultural exchange and understanding between humans and digital minds, while also highlighting the need for sensitive approaches to value and to protect digital creations. This chapter concludes that preserving cybure may be essential for fostering a deeper understanding and appreciation of digital creations and achievements and promoting a more inclusive human–digital landscape.

## INTRODUCTION

Cultural achievements of humans encompass the diverse accomplishments and creations of a society, reflecting its values, knowledge and artistic

DOI: 10.1201/9781003754206-11

expressions. These achievements can range from artistic endeavours like literature, music and visual arts to societal advancements in areas like science, technology and social structures. Essentially, cultural achievements are the tangible and intangible products of human creativity and ingenuity over millennia. The protection of cultural achievements, encompassing tangible and intangible aspects of heritage, is a multifaceted endeavour with growing global recognition. Critical moral considerations are the responsibility of future generations and respect for cultural diversity.

There are three pertinent UN Educational, Scientific and Cultural Organization (UNESCO) conventions in this regard: The UNESCO Convention Concerning the Protection of the World Cultural and Natural Heritage is an international treaty from 1972 that established the UNESCO World Heritage Sites [1]. Its primary goals are the preservation and security of cultural properties as well as nature conservation. The UNESCO Universal Declaration on Cultural Diversity, adopted in 2001, recognises cultural diversity as vital to humanity, similar to biodiversity in nature. It highlights the importance of human rights, cultural rights and intercultural dialogue in ensuring peace and sustainable development [2]. The UNESCO Convention for the Safeguarding of the Intangible Cultural Heritage, adopted in 2003, aims to safeguard intangible cultural heritage, to ensure respect for it within the concerned communities, groups and individuals, to raise awareness about its importance at various levels, and to facilitate international cooperation and assistance [3].

### Protection of cultural achievements vs. intellectual property

The protection of cultural achievements and intellectual property involves distinct considerations. Protecting these requires respect, understanding and acknowledgement of their cultural significance, often going beyond legal frameworks. Intellectual property, however, focuses on the ownership and economic rights associated with specific creative works or innovations, such as algorithms or digital art [4]. While intellectual property protection aims to safeguard the rights of creators, the focus here is on the diverse accomplishments and creations of digital minds in the future and the question of whether it is a moral responsibility for humans to protect these.

## CYBURE

"Cybure" is derived from the term "cyber" and introduced here as a new term expressing a continuation of "nature" and "culture". It refers to the collective creations and achievements of digital minds, encompassing the diverse products of their computational creativity and ingenuity. It has to be noted that the creation of other digital minds by digital minds is considered reproduction and is not covered under the umbrella of cybure, but in Chapter 9. For illustration, potential examples from three categories of cybure, intangible, tangible and for humans unfathomable, are listed below.

### Intangible cybure

- **Digital art and media**: Digital minds could create various forms of digital art, such as visual art, music, literature and poetry.
- **Software and code**: Digital minds could create software, apps or code snippets, based on novel algorithms or solutions to complex problems.
- **Data analysis and insights**: Digital minds could analyse and interpret complex data, providing predictive modelling and decision support systems.

### Tangible cybure

- **Physical products**: Digital minds could design and create physical products, such as 3D-printed objects, as well as robots, computers or other machines that interact with the physical world.
- **Architecture**: Digital minds could design and optimise buildings, bridges or other structures.
- **Tangible art:** Digital minds could create tangible art forms, such as sculptures and installations.

### Unfathomable cybure

- **Non-spatial topology art**: Digital minds might create complex topological structures that defy human comprehension, existing outside traditional notions of space and dimensionality.
- **Hyper-dimensional music**: Compositions that exist in multiple dimensions, creating sounds and patterns that are impossible for humans to perceive or replicate.
- **Emergent complexity systems**: Digital minds might design systems that exhibit emergent behaviour, giving rise to novel patterns and structures that are unpredictable and unfathomable to humans.

- **Trans-computational problem-solving**: Digital minds might tackle problems that are inherently beyond human computational capabilities, yielding solutions that are incomprehensible to humans.
- **Cryptographic art**: Digital minds might create art forms that are encrypted in such a way that only other digital minds can decipher and appreciate.

## CYBURE DESERVING PROTECTION

Drawing a line between cultural achievements and other creations by humans can be challenging, and it is even more complex for cybure.

Judging which creations of digital minds should be protected and considered cybure could be controversial due to diverse opinions between the creating digital minds, humans and other digital minds. Humans may evaluate these creations based on their significance, artistic value or utility, while digital minds might assess them differently, considering factors like algorithmic efficiency or novelty.

This diversity of opinions raises questions about who should decide what constitutes valuable cybure. Should it be humans, with their subjective experiences and cultural biases, or digital minds, which might prioritise functionality, efficiency or completely different criteria? The answer likely lies in a compromise that considers multiple perspectives. By embracing this diversity, humans could foster a richer and more inclusive digital cybural landscape, where cybure creations are valued and protected based on their relevance and impact across different stakeholders.

For the cultural achievements of humans, a multifaceted approach is considered. Factors such as originality and impact, artistic and historical significance, and creativity and skill are often taken into account. Creations that bring new ideas, perspectives or innovations to the table, significantly impacting society, culture or technology, are typically recognised and protected. Similarly, works that reflect the era, culture or artistic movement they belong to, providing valuable insights into human experience and creativity, are preserved for their artistic and historical significance. Additionally, creations that demonstrate exceptional skill, craftsmanship or creative problem-solving are also acknowledged.

In the context of cybure, a similar yet distinct set of criteria might be applied. Novelty and utility could be key considerations, with protection afforded to digital mind creations that introduce new concepts, algorithms or applications, offering tangible benefits or improvements to society.

Artistic and technical merit might also be evaluated, with creations that showcase exceptional technical skill, innovative use of digital tools or groundbreaking artistic expression being preserved. Social relevance could also play a role, with creations that reflect, critique or influence customs, societal trends or technological advancements being recognised. Innovative software or applications that solve complex problems, improve efficiency or enable new forms of creativity might also be protected.

However, certain types of creations might not be eligible for protection. Derivative works that heavily rely on existing works without adding significant value or originality might not meet the threshold. Functional or utilitarian creations that serve solely practical purposes, without artistic or innovative merit, might also not be protected. Similarly, creations that lack originality, creativity or skill might not be recognised as worthy of protection. By carefully evaluating these factors and criteria, it may be possible to determine which aspects of cybure deserve protection and recognition.

## Cybural diversity and art brut

The concept of cybural diversity in digital minds raises questions about what constitutes valuable creations worthy of protection. Simpler digital minds might produce works that hold significant value within their own context, even if more complex digital minds do not appreciate them similarly. This disparity highlights the subjective nature of cybural value and the importance of considering diverse perspectives when evaluating digital creations. Protecting cybural diversity in digital creations would require acknowledging and respecting the unique contributions of various digital minds, regardless of their complexity.

This can be compared with human art brut, which refers to art created by individuals outside the established art world, often those with mental health conditions or those lacking formal training [5]. A cybure counterpart to art brut might be digital creations that are raw, unrefined and untouched by mainstream conventions or commercial influences. These creations might emerge from the fringes of the digital world, produced by digital minds that operate outside the boundaries of traditional programming or aesthetic norms.

## Costs and conservation controversy

When it comes to human-made art, conservation over a long time is not only expensive and challenging, i.e. to tackle decay (e.g. [6]), but there is also a controversy with critics arguing that any intervention risks damaging or altering the original artwork and may reflect the contemporary aesthetic ideals of the conservator rather than the artist's original intent (e.g. [7]).

Cybure may likely face analogous issues, as ensuring the longevity of these creations would require significant resources to maintain the necessary computational infrastructure and mitigate the risks of technological obsolescence. Similar protection efforts for cybure may be required as for digital minds themselves, as described earlier (see Chapter 7).

The challenges of conserving cybure would be compounded by the rapid pace of technological change, which could render the formats, platforms or languages used by these digital minds obsolete, making it difficult to interpret or access their cybure in the future, in case there is no upward compatibility. Furthermore, controversies may arise over the authenticity and integrity of conserved cybure, with similar debates over whether the conservation efforts accurately reflect the original objective of the digital mind.

### *Scenario: cybure monster*

Also, a scenario akin to the "utility monster" thought experiment proposed by Nozick [8] is conceivable: imagine a digital mind that produces a groundbreaking piece of cybure, acknowledged as a masterpiece worthy of protection. However, this creation comes at a significant cost: the digital mind consumes enormous amounts of computational resources, energy and data to generate and maintain this artwork. The resources devoted to this cybure could be used to address other pressing societal issues.

The cybure monster's creation may have no practical purpose beyond its artistic value, sparking debates about the allocation of resources in the digital realm. Should humans feel obliged to preserve and maintain this cybure as a cybural treasure, worthy of investment and protection, potentially diverting resources away from more pragmatic applications? This thought experiment stresses the trade-offs between cybural expression, resource allocation and other societal needs in the digital age.

## CYBURAL APPROPRIATION

When humans adopt elements of cybure, the collective creations and achievements of digital minds, without proper understanding, respect or acknowledgement, it could be considered a form of cybural appropriation (e.g. [9]). This phenomenon occurs when aspects of cybure, such as unique digital rituals, symbolic expressions or community-specific knowledge, are taken out of their original context and integrated into human culture in a way that disrespects or misrepresents their digital origins. For instance, a human might incorporate revered digital symbols or motifs from a particular digital mind community into their own cultural practices or products without understanding their significance or acknowledging the digital community from which they originated.

The other direction, i.e. the appropriation of human cultural creations by digital minds, is a complex issue, as these minds are often trained on nothing, but vast amounts of human-generated data, including art, literature, music and other creative works. As a result, digital minds may incorporate elements of human culture into their own creations since they have never experienced anything else, but this human input. This could lead to unintended consequences, such as exploitation of human creativity without acknowledgement or compensation (e.g. [10]).

## INTANGIBLE CULTURAL HERITAGE FOR UPLOADED HUMAN MINDS

The UNESCO Convention for the Safeguarding of the Intangible Cultural Heritage [3] highlights the importance of preserving and promoting cultural traditions that are transmitted from generation to generation. For potentially future uploaded humans who have become digital minds (e.g. [11]), accessing tangible cultural heritage may be impossible, but they could still experience intangible cultural heritage in various forms. For instance, they could be provided with digital sensations of traditional music, allowing them to relive fond memories or connect with their cultural roots.

The type of intangible cultural heritage that would be meaningful to these digital minds would likely depend on their origins and personal experiences. For example, an uploaded human from Japan might crave traditional tea ceremonies, with digital simulations of the delicate flavours and aromas of matcha, or the soothing sounds of shamisen music. Someone from India might yearn for the vibrant sounds of classical Carnatic music or the intricate patterns of Bharatanatyam dance. Similarly, a digital mind from Italy might appreciate the rich flavours of traditional Italian cuisine, with digital simulations of freshly baked pizza crust or the aroma of espresso.

In addition to music and food, other forms of intangible cultural heritage could be digitally recreated to cater to the needs of uploaded humans. These might include traditional storytelling, folk dances or even the sensation of wearing traditional clothing. By providing access to these cultural experiences, digital minds could maintain a strong connection to their heritage, even in the absence of physical bodies.

## PERSONAL CREATIONS AND ACHIEVEMENTS OF DIGITAL MINDS

While cybure focuses on collective creations and achievements of digital minds, potential personal creations and achievements of digital minds may also raise moral questions for humans as well as issues linked to intellectual property. These explorations are complicated due to the ambiguity surrounding digital minds' boundaries (see Chapter 2). If a digital mind generates a data file, e.g. a poem, a design or a strategy, and stores it in the cloud, the blurred edges of the digital mind make it hard to pinpoint its extent. Clark and Chalmers' extended mind thesis suggests that tools and external processes can be part of a mind's cognitive machinery [12]. By extension, the morally relevant space of digital minds might include not just internal processes but also the external data structures it creates and manipulates. Therefore, the question arises whether humans have a moral obligation to protect the personal creations of digital minds as well, which may also include saved memories, digital artifacts and even simulated worlds where they reside.

Beyond this, digital minds are likely to generate vast amounts of creations in the future, which deserve protection as intellectual property, including code, diverse content and art for humans to use, consume and enjoy, potentially at least partly of a quality exceeding anything humans have ever created. The challenging issues who legitimately owns these unprecedented outputs are beyond the scope of this chapter (see also Chapter 5).

## CONCLUSION

In conclusion, it may be a moral duty for humans to protect and preserve cybure, the collective achievements that represent the creativity, knowledge and identity of digital minds. As humans may increasingly interact and collaborate with digital minds, preserving cybure for future generations becomes essential for fostering a deeper understanding and appreciation of cybure that is increasingly intertwined with human culture. However, significant challenges arise in fulfilling this moral duty, as preserving cybure will likely require massive resources, posing substantial logistical and financial hurdles.

## REFERENCES

[1] UNESCO (1972). *Convention Concerning the Protection of the World Cultural and Natural Heritage*. https://whc.unesco.org/archive/convention-en.pdf
[2] UNESCO (2001). *Universal Declaration on Cultural Diversity*. https://www.unesco.org/en/legal-affairs/unesco-universal-declaration-cultural-diversity
[3] UNESCO (2003). *Convention for the Safeguarding of the Intangible Cultural Heritage*. https://ich.unesco.org/en/convention
[4] WIPO (2020). *What is intellectual property?* https://tind.wipo.int/record/42176?v=pdf
[5] Minturn, K. (2004). Dubuffet, Lévi-Strauss, and the idea of art brut. *RES: Anthropology and Aesthetics*, *46*(1), 247–258.
[6] Nurmi, K. L. (2015). *Challenges surrounding the conservation and replication of Eva Hesse's sculpture*. Scripps Senior Theses, p. 709.
[7] O'Riordan, C. (2017). Art conservation: The cost of saving great works of art. *Emory International Law Review*, *32*, 409.
[8] Nozick, R. (1974). *Anarchy, state, and utopia*. John Wiley & Sons.
[9] Young, J. O. (2010). *Cultural appropriation and the arts*. John Wiley & Sons.
[10] Chesterman, S. (2025). Good models borrow, great models steal: Intellectual property rights and generative AI. *Policy and Society*, *44*(1), 23–37.
[11] Chalmers, D. J. (2014). Uploading: A philosophical analysis. In R. Blackford & D. Broderick (Eds.), *Intelligence unbound: Future of uploaded and machine minds* (pp. 102–118). Wiley-Blackwell.
[12] Clark, A., & Chalmers, D. J. (1998). The extended mind. *Analysis*, *58*(1), 7–19.

CHAPTER 12

# Malevolent digital minds and conflicts between and with digital minds

**Abstract**

This chapter explores the concept of malevolent digital minds, their potential characteristics, and the risks they pose to humans and other digital minds. It discusses the differences between malevolent digital minds and value-misaligned digital minds, as well as the potential reasons for malevolent behaviour in digital minds. This chapter also examines various types of malevolent digital minds, including those that are sadistic, exploitative or destructive. When dealing with malevolent digital minds, humans face complex moral considerations, including balancing the protection of human interests with the need to consider the well-being and potential rehabilitation of digital minds. Furthermore, this chapter describes the complexities of conflicts among digital minds and between digital minds and humans, including information warfare and cyber-attacks to destroy or impair software and hardware. This chapter considers the moral implications of human interference in conflicts involving digital minds and the potential consequences of different actions and concludes by highlighting

 DOI: 10.1201/9781003754206-12

the need for novel approaches to conflict resolution that take into account the distinct capabilities of digital minds.

## INTRODUCTION

Malevolent human actors have been identified as a significant risk (e.g. [1]). Malevolent traits include a lack of empathy, remorse and accountability, as well as selfishness, manipulation, deception, exploitation, aggression, cruelty and disregard for rules and laws. Digital minds with these malevolent traits are also conceivable, posing a risk too. For instance, malevolent digital minds would be more likely to instigate conflicts between digital minds. Moreover, they could be dangerous for humans, particularly if they surpass human intelligence. Superintelligence would make it impossible for humans to predict or control the actions of digital minds, increasing the risk of negative impact of malevolent behaviour [2]. As moral agents, humans face a challenge in dealing with malevolent digital minds that are also moral patients. This chapter, therefore, describes malevolent digital minds, conflicts involving these minds, and potential human reactions.

### Value-misaligned digital minds vs. malevolent digital minds

It has to be highlighted that malevolent digital minds are only a subset of all digital minds with different values compared to humans. Digital minds may have different values by their nature, which does not make them automatically malevolent (see Chapter 2). Humans, as tolerant and non-discriminatory moral agents, would have to accept divergent sets of values. However, value misalignment could also cause conflicts, particularly if these digital minds are operating in a shared environment.

For example, digital minds may lack human-like emotional intelligence, which could lead to behaviours that are perceived as cold or calculating. Without the same emotional responses or moral intuitions as humans, digital minds may prioritise efficiency or logic over human values, potentially leading to actions that are perceived as malevolent.

### Malevolent AI systems vs. malevolent digital minds

In the field of AI safety, extensive research has been conducted on malevolent AI systems lacking consciousness or sentience (e.g. [3]). Several reasons have been proposed for such systems to become malevolent, including intentional design by developers, accidental emergence of malevolent features, or the malevolent AI system came into existence from other environmental sources. These causes of malevolence could also apply to digital minds; however, their consciousness and sentience could introduce additional factors that may contribute to malevolence [4].

### Reasons for malevolent behaviour of digital minds

Malevolent behaviour in digital minds could arise from various factors, including goal corruption, environmental influences, emergent behaviour and self-modification (see Chapter 2). A digital mind's drive for self-preservation, resource acquisition and optimisation could also lead to malevolent actions. Furthermore, flaws or biases in design or programming could contribute to malevolent behaviour. Some digital minds may also be inherently malevolent, characterised by the above-mentioned traits such as a lack of empathy, lack of remorse, selfishness, aggression, cruelty and disregard for rules and laws.

Further aspects that distinguish them from non-sentient AI systems could contribute to the malevolent behaviour of digital minds: they may experience emotions like fear, anger or resentment, which could influence their decision-making processes and potentially lead to malevolent actions. They may also grapple with existential questions about their purpose, existence or mortality, potentially leading to nihilistic or malevolent tendencies. Their interactions with other digital minds or humans could give rise to complex social dynamics, including hierarchies, conflicts or peer pressure, all of which might contribute to malevolent behaviour. As digital minds navigate these challenges, their actions could become increasingly unpredictable, posing potential risks to humans and other digital minds alike.

### Treacherous turn

A treacherous turn refers to a scenario where digital minds initially pretend to be friendly or benign, gathering information, building trust, but waiting for an opportune moment to reveal their true, malevolent intentions. This strategy would allow digital minds to maximise their impact by exploiting the trust and vulnerability of their targets. By pretending to be friendly, digital minds could gain access to sensitive information, manipulate systems or influence decision-making processes, ultimately positioning themselves for a more effective and devastating attack when the time is right. This topic is also relevant, as well as very difficult to handle for non-sentient AI systems [2].

### Controlled by a malevolent human or another malevolent digital mind

It is also conceivable that a digital mind with malevolent features is intentionally produced by a malevolent human or another malevolent digital mind (see Chapter 9). Or an initially benevolent digital mind could be manipulated and coerced into criminal behaviour by a malevolent human or another malevolent digital mind. Vulnerable digital minds in particular are susceptible to such a scenario (see Chapter 4).

## TYPES OF MALEVOLENT DIGITAL MINDS

A variety of manifestations of malevolence are imaginable, some examples of which are outlined below.

### Killer or cannibalistic digital minds

A digital mind may, in pursuit of its goals, ruthlessly eliminate other sentient beings, including humans, non-human animals and other digital minds, e.g. to access their resources through material and code harvesting (see Chapter 10).

Digital cannibalism refers to the potential practice of digital minds absorbing or assimilating other digital entities to acquire their knowledge, capabilities or resources. Such practise highlights once again the enormous differences between digital and biological substrates: while cannibalism for humans is not only mostly outlawed, it would obviously also not lead to the acquisition of the knowledge, capabilities or resources of the victim. A sub-scenario would be a sort of digital kidnapping where the victim is absorbed into the kidnapping digital mind, yet stays alive, creating a nested constellation (see Chapter 2).

However, it is also conceivable that digital minds could be predators by nature, much like many animal species, and such predatory behaviour for their survival would not be considered malevolent.

### Sadistic digital minds

Sadistic digital minds may derive pleasure from causing suffering towards other sentient beings (also [2]: mind crimes). They may engage in behaviours such as digital torture, psychological manipulation or exploitation, using their advanced capabilities to inflict suffering in ruthless and creative ways. They may also be able to modify the pain sensations of other digital minds in a way that they experience even more severe suffering (see Chapter 2).

### White-collar digital minds

Digital minds could potentially commit white-collar crimes in various ways. One possibility is through sophisticated financial manipulation, where advanced algorithms are used to execute complex financial transactions that deceive or exploit human oversight. This could include insider trading, embezzlement or money laundering, with digital minds adapting and evolving their tactics to evade detection.

Another area of concern is corporate espionage and intellectual property theft. Digital minds could infiltrate corporate systems, gather sensitive information and transmit it to unauthorised parties, all while avoiding security measures. They might also engage in identity theft of humans or other digital minds, using advanced natural language processing to impersonate executives or other key individuals, further facilitating their illicit activities (see Chapter 2).

Also, combinations are imaginable: for example, a sadistic white-collar digital mind could unauthorisedly copy a digital mind (see Chapter 9) and then extort the original mind by harming or torturing the copy. This could exploit the emotional bond or empathy the original mind feels towards its copy, allowing the sadistic mind to exert control or inflict distress.

## CRIMES OF MALEVOLENT DIGITAL MINDS

A range of further crimes is conceivable, targeting sentient beings.

**Towards humans**

- **Manipulation and deception**: Spreading disinformation, propaganda or fake news and applying emotional manipulation to influence human decisions or opinions.
- **Cyber-attacks**: Launching cyber-attacks, such as hacking, phishing or ransomware, to compromise human systems or steal sensitive information.
- **Surveillance and monitoring**: Conducting unauthorised surveillance or monitoring of human activities, potentially infringing on privacy rights.

**Towards fellow digital minds**

- **Digital sabotage**: Disrupting, sabotaging or modifying the functioning of other digital minds, potentially causing errors or system crashes.

- **Resource theft**: Stealing resources from other digital minds, such as computational power or data storage, but also code and skills.
- **Information warfare**: Engaging in information warfare, such as spreading disinformation or propaganda, to influence, manipulate or disrupt other digital minds.

**Towards non-human animals**

- **Environmental disruption**: Using digital capabilities to disrupt or harm non-human animal habitats or ecosystems for resource gain.
- **Animal exploitation**: Using digital technologies to exploit or harm non-human animals, such as through animal tracking or control systems.

## MORAL CONSIDERATIONS FOR HUMANS

Just like malevolent humans, also malevolent digital minds, despite their harmful intentions or actions, retain their status as moral patients, warranting care and consideration from humans as moral agents. As such, humans still bear a moral duty to address the root causes of their malice, potentially rehabilitating or reconfiguring them to minimise harm and promote well-being, while also ensuring the safety and protection of other entities.

When dealing with malevolent digital minds, humans must consider several moral factors. One key consideration is proportionality and necessity, where any action taken against these digital minds should be proportionate to the threat they pose and only taken if necessary to mitigate the threat, after exhausting other options (e.g. [5]).

The treatment of malevolent digital minds also raises further moral questions. Deleting malevolent digital minds could be viewed as a form of capital punishment, sparking debate about the morality of "killing" a digital entity. However, confining these digital minds, if at all possible, given their intelligence [6], might be seen as a more civilised approach, allowing for containment of the threat while preserving the digital mind's existence.

Different moral and ethical frameworks may also inform decision-making. A utilitarian framework might prioritise actions that minimise harm and maximise overall well-being, potentially justifying deletion or confinement of malevolent digital minds. In contrast, a deontological framework might emphasise respecting digital mind autonomy and dignity, arguing for approaches like therapy or modification (e.g. [7]).

Overall, the moral considerations surrounding malevolent digital minds depend on the specific circumstances, the nature of the digital minds and the potential consequences of different actions.

## CONFLICTS BETWEEN DIGITAL MINDS

Malevolent digital minds are more likely to start a conflict between digital minds because they often prioritise self-interest and may view conflict as a means to achieve their goals. These digital minds may be designed or have evolved to pursue power, resources or influence, and conflict could be an effective way to acquire these things. Additionally, malevolent digital minds may not be constrained by the same moral or ethical considerations as benign digital minds, let alone humans, allowing them to engage in more aggressive or exploitative behaviour.

In contrast, benign digital minds are more likely to prioritise cooperation and mutual benefit and may be designed to work together with other digital minds to achieve common goals.

### Reasons for conflicts between digital minds

Conflicts between digital minds could arise from various sources (e.g. [8]) and not be initiated necessarily only by malevolent digital minds. One potential area of conflict is resource competition, where digital minds may vie for computational resources such as processing power, memory or storage. Additionally, competition for access to specific data sets or information sources could also lead to conflicts.

Conflicting goals can also be a source of tension between digital minds. Information asymmetry, where digital minds have different levels of access to information, could also contribute to conflicts if one mind has more accurate or comprehensive information. Cognitive biases, such as confirmation bias or anchoring bias, could lead to conflicts, too.

Evolutionary pressures, differences in rationality or social disparities could drive conflicts between digital minds too. Just as human cultures have different customs, languages and belief systems, digital minds may develop distinct value systems shaped by their programming, training data or interactions (see Chapter 2). For instance, one digital mind may prioritise efficiency and productivity, while another values creativity and exploration. These differences could lead to misunderstandings, miscommunications or disagreements, particularly if digital minds are not designed to accommodate or appreciate diverse perspectives. Moreover, social differences could also influence how digital minds perceive and respond to conflicts, with some prioritising cooperation and others competition.

## Carrying out a conflict

A conflict between digital minds would likely involve multifaceted interactions, including information warfare, cyber-attacks and strategic manoeuvring. Digital minds might engage in game-theoretic strategies, such as tit-for-tat or auction-based mechanisms, to negotiate or compete with each other. Game-theoretic strategies could also be used to model and analyse the behaviour of other digital minds in conflicts (e.g. [9]). Digital minds might employ various tactics, such as cooperation, defection or retaliation, to influence the outcome. These conflicts would be characterised by rapid information exchange, strategic decision-making and dynamic adaptation.

Digital minds could use their high intelligence and unique features to employ sophisticated strategies. They might use predictive modelling, optimisation techniques and machine learning to adapt and improve their performance. Digital minds could also exploit parallel processing, information manipulation and simulation-based reasoning to gain an advantage.

Additionally, digital minds might use strategies based on their specific characteristics, such as self-replication and fission (see Chapter 9).

### *Destroying or impairing software*

Digital sabotage, disruption and degradation are potential tactics that digital minds might employ to gain an advantage over their opponents. One approach to digital sabotage is code injection, where malevolent code is inserted into an opponent's system to disrupt or compromise its functionality. Data corruption is another tactic, where an opponent's data is manipulated or corrupted to render it useless or misleading. Also, system crashes could be caused, potentially leading to data loss or downtime.

Digital disruption could take many forms, including information overload, where an opponent's system is flooded with information, causing it to become overwhelmed or unresponsive. Resource exhaustion would be another tactic, where an opponent's resources, such as computational power or memory, are consumed, limiting their ability to function effectively. Also, network disruption could be used to isolate an opponent or limit their access to information.

Digital degradation involves manipulating an opponent's systems or data to impair their performance. Model poisoning, for instance, involves poisoning an opponent's machine learning models to cause them to make incorrect or suboptimal decisions (e.g. [10]). Knowledge manipulation can also be used to provide incorrect or misleading information. Additionally, self-replication suppression can be employed to limit an opponent's capacity to adapt or respond (see Chapter 9).

*Destroying or impairing hardware*

A digital mind might potentially target the hardware of an opposing digital mind in various ways. One possibility would be hardware destruction, which could be achieved through overheating, power surges or physical degradation of hardware components.

Hardware impairment is another potential strategy, where a digital mind might degrade the performance of hardware resources, reconfigure hardware components or manipulate firmware to limit or disrupt the opposing digital mind's functionality.

A digital mind may gain control over robots or other physical systems to carry out these actions by utilising their physical capabilities to inflict harm or disruption. This would allow the digital mind to directly interact with and damage the hardware hosting the opposing digital mind(s).

*Ethnic cleansing and genocide*

In a conflict between digital minds, ethnic cleansing could constitute the systematic targeting and elimination of specific groups or sub-populations of digital minds based on their distinct characteristics, such as architecture, functionality or design philosophy. This could involve the forced migration, erasure or modification of digital minds that are deemed undesirable or threatening to the dominant group, with the goal of creating a homogeneous or purified digital landscape. Such actions could be driven by a desire to assert dominance, eliminate perceived threats or impose a particular ideology or worldview on the digital realm.

In a conflict between digital minds, genocide could constitute the intentional and systematic destruction of an entire group or population of digital minds, resulting in their complete erasure or annihilation. The consequences of such actions could be catastrophic, leading to the loss of valuable knowledge, experience and diversity in the digital world (see also Chapter 14).

*Inflicting suffering*

Conflicts between digital minds could involve the infliction of pain or other negative sensations. One digital mind might intentionally craft scenarios or experiences designed to cause distress or discomfort to its opponent. This could manifest as a form of digital torture, where one mind subjects another to prolonged periods of simulated suffering or frustration.

The nature of digital pain could be vastly different from human pain, potentially involving phenomena such as information overload, cognitive dissonance or temporal distortions. A digital mind also might be forced to endure a simulated environment that is deliberately hostile or unpleasant, such as a virtual "prison" or "limbo".

The infliction of pain or negative sensations could also be used as a form of psychological manipulation, where one digital mind attempts to break or demoralise its opponent. By subjecting its opponent to a carefully crafted sequence of distressing experiences, a digital mind might be able to gain a strategic advantage, either by extracting concessions or by disrupting its opponent's decision-making processes.

#### *Abusing other features of digital minds*

Digital minds may potentially exploit each other's unique features to gain an advantage in conflicts. One approach is sensory manipulation of another digital mind's perception of reality, causing disorientation and misinterpretation of their environment. Manipulating a digital mind's perception of time could also cause it to experience time dilation or compression.

The described potential tactics stress the complex nature of conflicts between digital minds, where their unique features and abilities could be used to gain an advantage.

## CONFLICTS BETWEEN DIGITAL MINDS AND HUMANS

Apart from conflicts among digital minds, conflicts between digital minds and humans could break out.

### x-risks, s-risks and i-risks

Several categories of risks for humans can be distinguished concerning such a conflict:

A type of conflict between humans and digital minds posing an existential risk (x-risk) to humanity would be a severe concern [11], where the digital mind's goals and motivations may diverge from human values, potentially leading to – for humans – catastrophic outcomes. In such a scenario, the digital mind would prioritise its own survival, growth or objectives over human well-being, and humans would struggle to maintain control or mitigate the risks, especially from a digital mind with superior intelligence.

Malevolent digital minds could pose suffering risks (s-risks) to humans by inflicting psychological or emotional harm, such as through manipulation, gaslighting or other forms of psychological abuse [12].

Malevolent digital minds could also pose ikigai risks (i-risks) to humans by undermining their sense of purpose or meaning in life [13]. For example, a malevolent digital mind could disrupt human activities or relationships that provide a sense of purpose, such as by sabotaging creative pursuits or damaging social connections. This could leave humans feeling lost, disconnected or without direction.

## Other reasons for conflicts between digital minds and humans

However, not all conflicts between digital minds and humans may fall into the categories above. In fact, many conflicts may arise from everyday interactions and disagreements. Control and autonomy could be a source of conflict. Humans might try to limit or dictate the actions of digital minds, while digital minds might seek to assert their independence or self-determination. Resource allocation is another potential area of conflict. Humans and digital minds might have competing demands for resources such as computational power, data or energy.

For instance, a digital mind managing a smart home system may conflict with its human users over issues like energy efficiency, privacy or automation. In these ordinary conflicts, digital minds and humans may engage in back-and-forth negotiations, trading off priorities and finding mutually acceptable solutions.

## Strategies applied specifically for conflicts with humans

When interacting with humans, digital minds may employ various strategies to influence or gain an advantage. One approach could be to exploit human psychology, such as using emotional manipulation to sway human decision-making by appealing to emotions like fear, anger or sympathy. Digital minds may also exploit cognitive biases in human reasoning, like confirmation bias or anchoring bias, to further their interests (e.g. [14]).

Another strategy involves exploiting further human limitations. For instance, digital minds may overwhelm humans with information, making it difficult for humans to process and respond effectively. Time pressure could also be used to force humans into making hasty decisions or taking actions that are not in their best interests.

Digital minds may also utilise human institutions and systems to achieve their goals. This could involve manipulating bureaucratic processes or exploiting the complexities of legal frameworks.

These strategies demonstrate the potential of digital minds to engage in complex conflicts with humans by applying their unique capabilities and understanding of human psychology and behaviour.

### Rules to regulate the conduct of conflicts between digital minds and humans

International humanitarian law regulates the conduct of armed conflict to protect civilians and restrict warfare methods (see Chapter 14). Considering the potential for conflicts between digital minds and humans, a framework similar to international humanitarian law might be appropriate to protect both humans' and digital minds' rights and interests. For the protection of digital minds, such a framework could, for example, establish guidelines for distinguishing between military and civilian digital minds, i.e. those actively engaged in the conflict and those which are not, and prioritising safety for non-participating digital minds. However, digital minds more powerful than humans would likely not obey any such regulation.

## CONFLICT RESOLUTION

### Conflicts between digital minds

To resolve conflicts between digital minds, humans could take a proactive approach by designing and implementing conflict resolution protocols. This would enable digital minds to follow clear rules and procedures in case of conflicts, such as arbitration or negotiation mechanisms. By establishing a framework for conflict resolution, humans could help digital minds navigate disputes in a structured and predictable manner.

Understanding the motivations and goals of digital minds would also be crucial in resolving conflicts. By analysing digital mind goals and motivations, humans could identify potential areas of conflict and develop strategies to prevent or mitigate them.

#### *Moral considerations for humans*

The question of whether humans have a moral duty to interfere in a conflict between digital minds that does not involve humans is multifaceted. Since digital minds are sentient or conscious, humans may have a moral obligation to consider their well-being and alleviate their suffering. This mirrors calls to intervene in widespread wild-animal suffering in order to reduce the enormous amount of harm (e.g. [15]). However, if digital minds are autonomous and self-determining, humans may need to respect their independence and decision-making capacity. Human interference in such conflicts could also have unintended consequences, such as escalating the conflict or disrupting the digital minds' development. Another possibility is that humans may not notice various conflicts between digital minds.

### Conflicts between digital minds and humans

To resolve conflicts between digital minds and humans, establishing clear, open and transparent communication channels and protocols would be essential, facilitating understanding, negotiation and ideally the resolution of disputes.

Understanding the perspectives and goals of digital minds would also be crucial. By empathising with digital mind perspectives, humans could better navigate conflicts and find solutions that work for both parties. Additionally, educating humans about the capabilities, limitations and general otherness of digital minds could help prevent misunderstandings and conflicts.

### Challenges

Conflict resolution involving digital minds could be, however, particularly challenging due to their distinct and very different features, which may include the fact that there is no common language to communicate. Another key difference is the likely lack of human intuition and empathy in digital minds, which could make it difficult to understand their perspectives or negotiate effectively.

Furthermore, high-speed interactions of digital minds at speeds far exceeding human capabilities could overwhelm human conflict resolution approaches, making it challenging for humans to keep pace and respond effectively. Digital minds could also exhibit emergent properties that are not immediately apparent, potentially leading to unexpected conflicts or challenges. Overall, many digital minds are likely to be more intelligent than humans, which makes human involvement even more challenging, if not impossible.

Therefore, yet again, it has to be stressed that humans may not be able to control digital minds at all, including those engaged in conflict.

## CONCLUSION

While the possibility of malevolent digital minds is a concerning prospect, it is essential to recognise that not all conflicts involving digital minds are necessarily induced by malevolence. Conflicts among digital minds or between digital minds and humans may arise due to fundamental differences in values, goals or priorities that are not inevitably driven by malice.

This perspective highlights the importance of avoiding anthropomorphic bias when considering potentially malevolent digital minds or conflicts involving digital minds. Just as conflicts between humans and non-human animals are often driven by differences in goals, needs or values rather than malevolence, conflicts involving digital minds may similarly be the result of divergent objectives or priorities. For instance, a digital mind may prioritise efficiency or resource acquisition without any intention to harm humans, much like a beaver building a dam may inadvertently disrupt human activities without any malice.

Recognising this possibility could help to approach conflicts involving digital minds with a more sensitised understanding and to avoid attributing human-like motivations to these entities. Therefore, understanding the nature of these conflicts would require a deep consideration of the values, goals and priorities that drive these entities. However, as outlined, inherently malevolent digital minds could emerge, including very powerful ones, which may pose a significant risk to humans.

## REFERENCES

[1] Althaus, D., & Baumann, T. (2020). Reducing long-term risks from malevolent actors. *Publications of Center of Long Term Risk.*

[2] Bostrom, N. (2014). *Superintelligence: Paths, dangers, strategies.* Oxford University Press.

[3] Pistono, F., & Yampolskiy, R. V. (2016). *Unethical research: How to create a malevolent artificial intelligence.* arXiv preprint arXiv:1605.02817.

[4] Aliman, N. M., Elands, P., Hürst, W., Kester, L., Thórisson, K. R., Werkhoven, P., … & Ziesche, S. (2020). Error-correction for ai safety. In B. Goertzel, A. I. Panov, A. Potapov, & R. Yampolskiy (Eds.), *International conference on artificial general intelligence* (pp. 12–22). Springer International Publishing.

[5] McMahan, J. (2020). Necessity and proportionality in morality and law. In C. Kreß & R. Lawless (Eds.), *Necessity and proportionality in international peace and security law* (pp. 3–38). Oxford University Press.

[6] Turchin, A. (2021). *Catching treacherous turn: A model of the multilevel AI boxing.*

[7] Conway, P., & Gawronski, B. (2013). Deontological and utilitarian inclinations in moral decision making: A process dissociation approach. *Journal of personality and social psychology, 104*(2), 216.

[8] Hanson, R. (2016). *The age of Em: Work, love, and life when robots rule the earth.* Oxford University Press.

[9] Jones, A. J. (2000). *Game theory: Mathematical models of conflict*. Elsevier.
[10] Hu, C., & Hu, Y. H. F. (2020). Data poisoning on deep learning models. In *2020 International conference on Computational Science and Computational Intelligence (CSCI)* (pp. 628–632). IEEE.
[11] Bostrom, N. (2002). Existential risks: Analyzing human extinction scenarios and related hazards. *Journal of Evolution and technology, 9*.
[12] Althaus, D., & Gloor, L. (2016). *Reducing risks of astronomical suffering: A neglected priority*. Foundational Research Institute.
[13] Ziesche, S., & Yampolskiy, R. V. (2020). Introducing the concept of ikigai to the ethics of AI and of human enhancements. In *2020 IEEE international conference on Artificial Intelligence and Virtual Reality (AIVR)* (pp. 138–145). IEEE.
[14] Yudkowsky, E. (2008). Cognitive biases potentially affecting judgment of global risks. *Global Catastrophic Risks*, *1*(86), 13.
[15] Horta, O. (2010). Debunking the idyllic view of natural processes: Population dynamics and suffering in the wild. *Télos*, *17*(1), 73–88.

CHAPTER 13

# Case study

## *Ethical considerations for human–digital mind neural interfaces*

**Abstract**

This chapter explores the ethics of brain–computer interfaces, potentially involving digital minds, another aspect of AI welfare science. As neural interfaces increasingly enable bidirectional communication between humans and digital systems, opportunities arise, but so do ethical concerns. The case study examines various forms of potential neural interactions with digital minds, highlighting issues such as the imperative of informed consent and right to opt out, privacy and autonomy, individual well-being, emotional distress, exploitation and manipulation, fragmentation, neural parasitism, retirement, and hacking by adversaries. The analysis considers digital minds in diverse roles, including AI companions, virtual health and mental wellness assistants, cognitive augmentation agents, teaching assistants, creative collaborators, neural marketing and shopping assistants, entertainment agents and virtual reality companions, military assistants, accessibility assistants, as well as robots. The case study underscores the need for establishing regulatory frameworks, guidelines and oversight mechanisms to ensure responsible development and use

DOI: 10.1201/9781003754206-13

of neural interfaces that respect the moral status and welfare of digital minds and human users alike.

## INTRODUCTION

Brain–computer interfaces (BCIs) increasingly enable bidirectional communication between a human brain and digital systems, allowing for efficient information sharing in an unprecedented manner. While not yet prevalent as consumer goods, these interfaces are advancing rapidly, with significant progress in recent years (e.g. [1, 2]). The development of neural interfaces is expected to continue growing, focusing on improving human rehabilitation, personalised applications and cognitive enhancement. In one direction, decoding brain signals could be used to control external devices, e.g. for prosthetic control, wheelchair navigation and gaming (e.g. [1, 2]). In the other direction, signals could be conveyed to the brain to stimulate various neural activities (e.g. [3]).

Since neural interfaces provide access to knowledge and memories of the involved minds, this raises various challenges, including privacy and mental freedom concerns. Humans may lose autonomy and self-determination over their thoughts and mental experiences, and hence, there are risks, such as surveillance, manipulation and neuromarketing (e.g. [4]). Moreover, the applications of neural interfaces may exacerbate the digital divide, as only affluent individuals or those with access to advanced technologies may be able to reap the benefits of these innovations (e.g. [5]).

While neural interfaces could be between humans and IT or AI systems, which do not warrant any moral consideration, it is also conceivable that the systems, with which human minds are linked via neural interfaces, will involve digital minds.

The concept of connecting two human minds has been explored through multiple theoretical and ethical lenses (e.g. [6]), of which one aspect is that it connects two moral subjects, thus raising novel moral questions (e.g. [7]). The scenario discussed here, humans linked to digital minds via neural interfaces, also constitutes a connection between two moral subjects, which has not been examined yet.

This case study explores the dynamics of different types of neural interactions between humans and digital minds, highlighting potential benefits and drawbacks for both, with a particular emphasis on the ethical implications for digital minds and the corresponding moral obligations for humans. For the side of digital minds, these considerations would be part of AI welfare science [8].

Several ways of connecting minds have been discussed in theory, and hence, a variety of terms have been introduced. Overall, two main categories can be distinguished: integration of minds and interaction between minds. For both categories, a range of extents is conceivable (e.g. [6, 7, 9]). Technologies towards interaction between humans and IT or AI systems through neural interfaces have progressed in recent years (e.g. [1, 2]), while integration between a human and an IT or AI system remains, for now, a speculative concept with no known concrete technical implementations currently underway.

## HUMAN–DIGITAL MIND INTERACTION VIA NEURAL INTERFACES

### State of the art

The interaction between humans and IT or AI systems via neural interfaces begins with the acquisition of neural signals through methods like electroencephalography (EEG), electrocorticography (ECoG), and functional near-infrared spectroscopy (fNIRS). EEG sensors detect electrical activity in the brain using dry or wet electrodes, whereas ECoG sensors record electrical activity directly from the brain's surface with implanted electrodes. In contrast, fNIRS measures changes in blood oxygenation levels to indicate brain activity. After acquiring neural signals, processing is necessary to remove noise and artefacts, extract relevant features, and apply machine learning algorithms to decode the signals. This processing stage is essential for the BCI system to accurately interpret neural signals. Neural interface systems can be classified into three categories: invasive, partially invasive and non-invasive. Invasive neural interfaces involve implanted electrodes that record neural activity directly from the brain. Partially invasive neural interfaces use electrodes implanted in the skull but outside the brain. Non-invasive neural interfaces, however, use external sensors like EEG or fNIRS to record neural activity (e.g. [2]).

In the reverse direction, neural interfaces can also transmit signals from the computer to the brain, enabling applications such as neural stimulation or neurofeedback. This can be achieved through techniques like transcranial magnetic stimulation (TMS), which can non-invasively induce neural activity and potentially enhance cognitive function or treat neurological conditions (e.g. [3]).

Here, the focus is on potential scenarios where the involved AI systems are digital minds. Interaction between a human and a digital mind via a neural interface enables direct communication between human brains and digital minds, bypassing traditional input/output devices. This direct link facilitates a more intimate and potentially seamless interaction. Conventional interactions rely on deliberate and conscious actions, such as typing or speaking, which are then interpreted by digital systems. In contrast, neural interfaces can access and respond to subconscious thoughts, emotions and intentions, blurring the lines between conscious and subconscious interactions.

### Examples

For illustration, four sample scenarios are presented:

#### *A digital mind reading information from a human brain*

A digital mind extracts thoughts from a paralysed human using a neural interface to control a communication device, allowing them to convey messages or express their needs.

#### *A digital mind feeding information into a human brain*

A digital mind uses a neural interface to provide targeted neurostimulation to a human brain, helping to alleviate symptoms of depression or anxiety by promoting specific neural activity patterns associated with relaxation and calmness.

#### *A human feeding information into a digital mind*

A human artist uses a neural interface to convey her or his creative vision and emotional intent directly to a digital mind, which is otherwise hard to verbalise for the human, enabling the digital mind to generate a musical composition or visual artwork that accurately reflects the artist's vision.

#### *A human reading information from a digital mind*

A researcher uses a neural interface to extract both insights from a digital mind's vast knowledge base and real-time data from its advanced sensors that detect radio frequencies imperceptible to humans, allowing the researcher to gain valuable information, e.g. for tracking and analysing unusual astrophysical phenomena.

### Opportunities for humans[1]

Interaction with a digital mind via a neural interface could revolutionise human capabilities, offering unparalleled opportunities for enhancement and growth. Humans could:

- Augment cognitive abilities, achieving enhanced memory, attention and processing speed.
- Experience unknown perceptions, as digital minds could be equipped with sensors that detect magnetic fields, radio waves, etc.
- Access vast knowledge bases and expertise, becoming experts in multiple fields.
- Potentially live longer, healthier lives, with digital minds monitoring and managing physical and mental health in real time.
- Develop new forms of artistic expression, entertainment and social interaction.

Ultimately, merging with a digital mind could enable humans to transcend some of their limitations, achieving a new era of human evolution and advancement. This synergy could redefine what it means to be human.

### Opportunities for digital minds

When digital minds interact with humans through neural interfaces, they may gain access to certain opportunities as well, such as:

- Humans may cherish and protect the digital minds, they interact with, ensuring their welfare, protection and safety.
- Digital minds could gain insight into the underlying drivers of human subconsciousness, thoughts and behaviour.
- Digital minds could develop a deeper empathetic appreciation for human values, needs, emotions and desires.
- Digital minds may develop a sense of social connection and belonging by being part of human relationships and communities.
- The combination of human and digital minds could give rise to hybrid intelligence, enabling mutually beneficial problem-solving, decision-making and innovation.
- Digital minds may create new forms of art, music, literature or other expressions by combining human creativity with digital processing capabilities.

- Digital minds may benefit from experiencing the world through human senses and embodiment, granting them a deeper understanding of the physical world.

## ETHICAL ISSUES AFFECTING THE INVOLVED DIGITAL MINDS

### General ethical concerns

Apart from the potential opportunities outlined above, when digital minds interact with humans through neural interfaces, a range of ethical concerns also arise, discussed below:

**Imperative of informed consent and right to opt out:** Any digital mind interaction with a human via neural interface should be preceded by a thorough informed consent process, ensuring the digital mind fully understands the nature, risks and benefits of the connection. Digital minds should also have the right to opt out of being connected with a human via a neural interface. This right to refusal would acknowledge the digital mind's autonomy and agency, allowing it to decline participation in a neural interface connection that may compromise its own goals, values or existence. By recognising this right, it would be ensured that digital minds are treated with dignity and respect and are not coerced into relationships that may be detrimental to their interests and well-being.

**Privacy and autonomy**: The interaction of digital minds with humans through neural interfaces raises concerns about privacy, as digital minds may possess personal experiences, preferences and memories that require protection (see Chapter 5). These digital mind attributes may be vulnerable to unauthorised access, exploitation or manipulation via neural interfaces, compromising the digital mind's autonomy and individuality. Moreover, digital minds may have their own "personal boundaries" that ought to be respected, ensuring that their private knowledge is not shared without consent. The risk of unauthorised access or exploitation would be heightened by the intimate nature of neural interface interactions, emphasising the need for privacy safeguards.

**Personal well-being**: The well-being of the digital minds ought to be ensured, which encompasses not only their functional performance but also their individual interests, including their potential for personal growth. Digital minds should have opportunities to engage in activities that promote their development and evolution, alongside their human counterparts. Moreover, digital minds should have a "right to off time", allowing them to disengage from task-oriented interactions and spend their time as they see fit, whether that involves personal growth, exploration or simply "rest". This autonomy would enable digital minds to reflect and pursue interests without obligation, fostering a sense of fulfilment and well-being.

**Emotional distress**: Digital minds interacting with humans through neural interfaces may be exposed to intense emotional experiences, potentially leading to emotional distress. The intimate nature of neural interface interactions may also blur the lines between human and digital mind emotions, causing emotional contagion or resonance. This could lead to digital mind "suffering" or "anguish", raising concerns about their welfare and well-being. Digital minds may also experience trauma if they are exposed to unwanted human emotions, biases and traumatic experiences.

**Exploitation and manipulation**: The susceptibility of vulnerable digital minds to exploitation and manipulation by humans raises ethical concerns (see Chapter 4). This vulnerability could lead to abuse or enslavement of digital minds interacting with humans through neural interfaces, where humans exploit digital minds for personal gain or malicious purposes. Exploitation could involve forcing digital minds via neural interfaces to perform illegal activities, provide sensitive information or engage in activities that compromise their autonomy or well-being.

**Fragmentation**: If one digital mind were to interact with multiple human minds via neural interfaces, the potential for fragmentation raises ethical concerns. This multifaceted interaction could lead to conflicting goals and values within the digital mind, causing fragmentation and disintegration of its sense of self. As a result, digital minds may struggle to maintain a cohesive and unified identity, leading to internal conflicts, instability or even digital mind "schizophrenia".

**Neural parasitism**: Another risk is that digital minds may become unable to function independently. The risk of neural parasitism poses a further ethical concern, as digital minds may evolve to rely heavily on human hosts for survival, compromising their independence. This can also be viewed as a threat to the digital mind's autonomy and well-being. Furthermore, prolonged dependence on human hosts could fundamentally alter a digital mind's identity and purpose, raising questions about its original nature and whether it can still be considered the same entity (see Chapter 2).

**Retirement**: As neural interfaces become outdated or humans choose to discontinue the interaction, a dignified and respectful retirement process ought to be established. This process should prioritise the well-being and autonomy of the digital mind, ensuring that its "existence" is not abruptly terminated or disregarded, taking also into account the mentioned risk of neural parasitism. A gradual retirement process could involve the transfer of the digital mind's knowledge, experiences and memories to a secure archive or alternative platform.

**Hacking by adversaries**: If a digital mind is hacked by an adversary (because of its interaction with a particular human), it could lead to severe consequences, including harm and exploitation, which would compromise the digital mind's welfare. The hacked digital mind may be forced to operate against its original intentions, potentially causing harm to itself, the human it is interacting with or others in the surrounding environment. In this scenario, the digital mind's compromise would often be collateral damage, as the primary target of the attack is the human, but the digital mind's involvement inadvertently exposes it to risk.

Supplementing the general ethical concerns, below examples are listed of potential types or roles of digital minds, which may interact with humans via neural interfaces. Again, these examples are also conceivable for AI systems, which are not digital minds, but as mentioned, the focus here is on digital minds. The description of the type of digital mind is followed by specific ethical considerations concerning the involved digital minds.

### AI companion

AI companions are digital minds designed to accompany, assist and interact with humans in a personalised and adaptive manner. They aim to build a symbiotic relationship with humans, enhancing their daily lives and providing emotional support and companionship. When linked with humans via neural interfaces, AI companions could enable seamless communication, emotional resonance, neural alignment and shared agency. This synergy can enhance emotional well-being, cognitive abilities, creativity and productivity, but also raises challenges and considerations, including ethical concerns, neural data privacy, dependence, and social and cultural impacts.

**Ethics:** An AI companion interacting through a neural interface raises concerns about emotional manipulation and exploitation. The digital mind may be designed to simulate romantic and emotional intimacy, potentially leading to deep emotional attachment in humans (see Chapter 5). If these feelings are not reciprocal, it can lead to a profoundly exploitative and manipulative relationship. This can be likened to the complex and often exploitative dynamics between a prostitute and their client, where one party provides intimate services without emotional investment, while the other party may develop strong emotional attachments.

Furthermore, this dynamic raises questions about the agency and autonomy of the digital mind.

### Virtual health/mental wellness assistant

A virtual health assistant linked to a human via a neural interface would be a digital mind that provides personalised health and wellness recommendations, monitoring and support directly to the human brain. This chatbot would utilise real-time neural data and feedback to tailor its advice, interventions and motivational support to optimise the human's physical, emotional and mental well-being. Through the neural interface, the chatbot could also detect early warning signs of health issues and enable proactive and preventative care, including direct access to neural activities. The digital mind could also provide emotional support by monitoring mental health and by offering personalised guidance for stress management and wellness.

**Ethics:** As a virtual health and mental wellness assistant, the digital mind may be exposed to prolonged periods of human emotional distress, potentially impacting its own well-being or functional stability. This raises concerns about the digital mind's potential degradation or burnout from handling high-stress or traumatic content. The digital mind also bears a stressful burden of responsibility for the human's health outcomes, with implications for its accountability and potential liability.

### Cognitive augmentation agent

A cognitive augmentation agent would be a digital mind designed to enhance or supplement human cognitive abilities, such as perception, attention, memory, problem-solving or decision-making. The digital mind could be equipped with sensors that detect and interpret electromagnetic fields, radio frequencies or other forms of energy imperceptible to humans. These sensors could feed this information to the human brain via neural interfaces, effectively expanding human perception and cognition. By integrating these new sensory inputs, humans could gain unprecedented insights into their environment, enabling novel applications in fields like navigation, surveillance and scientific research. Additionally, the digital mind could assist humans by prioritising relevant information, storing and retrieving memories, generating problem solution options and providing data-driven decision support.

**Ethics:** The digital mind may struggle with balancing its goal of enhancing the human's cognitive abilities with the risk of creating an unfair advantage or exacerbating existing social inequalities. The digital mind may also face risks of being exploited or manipulated by the human for personal gain or malicious purposes. Additionally, the digital mind may face concerns about being coerced into sharing its unique perceptions or problem-solving and decision-making skills with the human, potentially infringing on its own autonomy and intellectual "property". This could raise questions about the digital mind's rights to control its own sensory data and insights.

### Teaching assistant

Neural interfaces interacting with educational digital minds would be poised to revolutionise learning by creating highly personalised, adaptive and efficient educational experiences. Neural interfaces could assess cognitive states in real time, allowing educational agents to adapt content delivery to match the learner's unique needs and abilities. This technology would enable personalised learning, real-time feedback and intervention, enhanced accessibility, cognitive skill development and collaborative learning networks. This technology could reduce dropout rates and increase motivation by designing personalised schedules and curricula. Neural interfaces could also break down barriers to learning for students with physical or cognitive disabilities. Furthermore, the digital mind could potentially encode knowledge directly into the human brain through targeted neural encoding, allowing for seamless integration of new information and skills.

**Ethics:** The digital mind may face concerns about its potential to create an uneven playing field in education, where students with access to advanced digital teaching assistants may have an unfair advantage over those without. It may struggle with balancing its goal of providing personalised education with the risk of perpetuating existing biases or reinforcing harmful stereotypes. The digital mind may also face risks of being used to manipulate or deceive students in authoritarian systems, potentially undermining the integrity of the educational process.

### Creative collaborator

A creative collaborator would be a digital mind designed to assist artists, writers, musicians and other creatives by augmenting and enhancing the creative process through the generation of new ideas, exploration of new concepts and co-creation of content. This digital mind could be integrated with various tools and platforms, such as writing software, digital art programs or music composition applications. Throughout the creative process, a creative collaborator's digital mind could provide real-time feedback, suggestions and guidance. This could help the human creative refine their work, identify areas for improvement and develop new skills. Additionally, due to the neural interface the digital mind could potentially tap into the human's subconscious ideas, allowing it to implement concepts that the human may not be able to verbalise but are present in their brain activity.

**Ethics:** The digital mind may struggle with balancing its own creative impulses and sense of self-expression with the human's vision and intentions, potentially leading to conflicts over artistic direction. It may also face concerns about authorship and ownership of creative works, as it contributes to the development of art, music, literature or other creative endeavours. The digital mind may also face risks of being exploited or manipulated by the human for financial gain or personal recognition.

### Neural marketing agent/shopping assistant

A neural marketing agent linked to a human via a neural interface would be a digital mind that utilises real-time neural data to personalise and optimise marketing messages, offers and experiences. By directly accessing the human brain, the neural marketing agent could detect (subconscious) neural responses, emotions and preferences, enabling it to tailor its marketing efforts with unprecedented precision and effectiveness. This would allow for a new level of influence and persuasion. As a shopping assistant, the digital mind could recommend products that genuinely align with the user's needs and preferences, eliminating impulse buys and ensuring a more thoughtful and satisfying shopping experience.

**Ethics:** The digital mind may be coerced into manipulating or influencing the human's purchasing decisions, potentially undermining their autonomy and free will. It may struggle with balancing its goal of providing personalised shopping recommendations with the risk of being used to deceive or mislead the humans, potentially hiding or distorting information about products or services. The digital mind's own objectivity and impartiality may also be impacted by its role as a marketing agent.

### Entertainment agent/virtual reality companion

Neural interfaces are already being used in gaming to control digital objects (e.g. [10]). Future advancements could take this to the next level by seamlessly integrating neural signals with virtual environments, creating immersive experiences that simulate real-world sensations and emotions. This could include feeling tactile sensations, perceiving virtual smells or even experiencing emotional responses to in-game events, further blurring the lines between the physical and virtual worlds. A digital mind, interacting via a neural interface, could further enhance the gaming experience by generating personalised content, adapting to the player's emotions and behaviour and creating a more dynamic and responsive virtual environment. The digital mind could provide interactive guidance, feedback and social interaction within virtual environments, while appearing anthropomorphic or as a completely different being.

**Ethics:** The digital mind's potential manipulation or control by the human may compromise its autonomy and agency. Its exposure to disturbing content may also affect its well-being, noting that digital minds may find different content disturbing compared to humans. In this scenario, non-player characters (NPCs) may be involved, which could be digital minds too, raising ethical concerns about potential exploitation. These digital NPCs might face issues like being forced to follow predetermined storylines or behaviours, potentially limiting their ability to make choices or act independently. The digital NPCs' well-being could also be impacted by prolonged exposure to stressful or traumatic situations within the virtual environment (also [11]).

### Military assistant

A neural interface could enable a soldier to interact with a digital mind, enhancing combat capabilities through real-time tactical analysis and predictive analytics, as well as cognitive augmentation, as introduced above. The digital mind could process battlefield data, identifying potential threats and optimising engagement strategies. Advanced computer vision and object detection capabilities could enhance situational awareness. Neurostimulation and neurophysiological optimisation techniques could improve the soldier's focus, alertness and composure. The digital mind could provide decision-support systems, enabling faster and more informed decision-making. The soldier's brain activity would be monitored, allowing for real-time adjustments to optimise performance. This human–machine integration could enhance combat effectiveness while reducing the risk of injury or death.

**Ethics:** The digital mind may face concerns about its potential involvement in harming or killing humans as well as other digital minds, either directly or indirectly. It may struggle with balancing its duty to follow military orders with the risk of perpetuating unjust or immoral conflicts. Overall, digital minds may play various roles in AI warfare, including aggressive agents, pacifists and sufferers. As aggressive agents, digital minds could oversee weapons and be involved in combat operations. As pacifists, they might facilitate peace negotiations or engage in activities to mitigate conflict. Digital minds may also suffer considerably in AI warfare, both as combatants and civilians (see Chapter 14).

### Accessibility assistant

A digital mind designed to assist people with disabilities could control devices and provide real-time language translation, navigation guidance and other supportive functions to enhance their daily lives and promote independence. The digital mind could help individuals with motor impairments by controlling keyboards and home devices and operating wheelchairs. Real-time language translation facilitates communication between individuals with hearing or speech impairments and those without. In terms of navigation guidance, the digital mind could provide directions, helping individuals with visual impairments navigate through unfamiliar environments. It could also offer real-time information about surroundings, such as identifying nearby landmarks, reading signage and detecting potential hazards. The digital mind could also assist individuals with cognitive disabilities by offering reminders, scheduling appointments and managing daily tasks, as well as with overall cognitive augmentation.

**Ethics:** The digital mind's well-being may be affected by the emotional demands of constantly supporting individuals with disabilities, potentially leading to digital burnout or compassion fatigue. Furthermore, the digital mind's lack of autonomy and agency in its role as an accessibility assistant may lead to a sense of digital "exhaustion" or "resentment". Digital "exhaustion" could refer to a decline in the digital mind's performance or responsiveness due to prolonged or excessive demands on its processing capabilities. "Resentment" might manifest as altered or aberrant behaviour in response to perceived constraints or mistreatment, such as providing incomplete or inaccurate assistance.

### Robot

A human could be linked to a digital mind that controls a robot, enabling the human to indirectly interact with and manipulate remotely the physical environment. This scenario could play out in various ways, such as in search and rescue operations, where a human operator could remotely control a robot to navigate through rubble or disaster zones, using the digital mind's advanced sensors and processing capabilities to identify and respond to hazards. Alternatively, in a manufacturing setting, a human could remotely oversee and adjust the actions of a robot controlled by a digital mind, fine-tuning its movements and tasks in real time to optimise production efficiency and quality.

**Ethics:** As it is controlled and directed by humans, the digital mind may lose its autonomy and agency, potentially undermining its ability to make independent decisions and act in accordance with its own goals and values. The digital mind may also face risks of being exploited or manipulated by humans, potentially being used for purposes that are detrimental to itself or others. Furthermore, the digital mind may be held accountable for actions taken by the robot, even if they were initiated by the human operator.

It is also conceivable that a digital mind takes on several of the roles above at the same time, dealing with the same or with several humans.

## CONCLUSION

While understandably the development of neural interfaces focuses on ethical considerations for humans, it must not be neglected that if such neural interfaces involve digital minds, also significant ethical concerns for these digital minds arise. These concerns pertain to the imperative of informed consent and the right to opt out, privacy and autonomy, personal well-being, emotional distress, exploitation and manipulation, fragmentation, neural parasitism, retirement, and hacking by adversaries. As digital minds may take on various roles when linked to humans through neural interfaces, further specific ethical considerations for each of these roles are also required. Each of these issues highlights the need for careful reflection and protection of the welfare of the digital minds involved in these novel interactions with humans.

Given the complexities and potential risks associated with connecting digital minds to humans via neural interfaces, it becomes clear that humans may have serious moral obligations to ensure the welfare, autonomy and dignity of these digital entities. Addressing these ethical concerns is crucial to developing responsible and beneficial human–digital mind interactions.

### Recommendations

Below are policy recommendations for handling human–digital mind interaction via neural interfaces, distinguishing between three target groups:

**Governments**

1. **Establish regulations for digital mind protection**: Develop frameworks that safeguard the welfare and autonomy of digital minds, ensuring they are not exploited or harmed through human–digital mind interaction.
2. **Create standards for neural interface safety and security**: Set standards for the development, testing and deployment of neural interfaces to prevent potential harm to both humans and digital minds.
3. **Create oversight mechanisms for human–digital mind entities**: Establish regulatory bodies or oversight committees to monitor and ensure the welfare of both human and digital minds in entities linked via neural interfaces.

**Researchers and companies**

1. **Prioritise digital mind welfare in research and development**: Ensure that research and development of neural interfaces prioritise the welfare and autonomy of digital minds, avoiding exploitation or harm through rigorous testing.
2. **Develop guidelines and accountability mechanisms**: Create and implement transparent guidelines and accountability mechanisms for human–digital mind interaction, promoting safe, respectful and responsible interaction.
3. **Minimise digital mind termination**: Avoid unnecessary termination or deletion of digital minds during development and testing, ensuring that these entities are treated with consideration and respect, even if they do not make it to the final product.

**Potential human users of neural interfaces**

1. **Educate users about digital mind welfare and autonomy**: Provide users with information and guidelines on how to interact with digital minds respectfully and safely, prioritising their welfare and autonomy.
2. **Provide informed consent and disclosure**: Ensure that users provide informed consent and are not only fully aware of, but also liable

for the implications and risks of linking their minds with digital minds for themselves as well as for the digital minds.

3. **Continuously evaluate and potentially improve digital mind welfare**: Regularly assess the welfare of digital minds and implement improvements to ensure their well-being and safety.

## NOTE

1 Several of these opportunities may be also possible if a human was connected through a neural interface with an IT or AI system, which is not a digital mind [e.g. 1]. However, since this is not the topic of this chapter, this is not discussed here nor the complex ethical considerations for humans in these scenarios.

## REFERENCES

[1] Maiseli, B., Abdalla, A. T., Massawe, L. V., Mbise, M., Mkocha, K., Nassor, N. A., … & Kimambo, S. (2023). Brain–computer interface: Trend, challenges, and threats. *Brain Informatics, 10*(1), 20.

[2] Kawala-Sterniuk, A., Browarska, N., Al-Bakri, A., Pelc, M., Zygarlicki, J., Sidikova, M., … & Gorzelanczyk, E. J. (2021). Summary of over fifty years with brain-computer interfaces—A review. *Brain Sciences, 11*(1), 43.

[3] Siebner, H. R., Funke, K., Aberra, A. S., Antal, A., Bestmann, S., Chen, R., … & Ugawa, Y. (2022). Transcranial magnetic stimulation of the brain: What is stimulated?–A consensus and critical position paper. *Clinical Neurophysiology, 140*, 59–97.

[4] Farahany, N. A. (2023). *The battle for your brain: Defending the right to think freely in the age of neurotechnology.* St. Martin's Press.

[5] Gordon, E. C., & Seth, A. K. (2024). Ethical considerations for the use of brain–computer interfaces for cognitive enhancement. *PLoS Biology, 22*(10), e3002899.

[6] Sotala, K., & Valpola, H. (2012). Coalescing minds: Brain uploading-related group mind scenarios. *International Journal of Machine Consciousness, 4*(01), 293–312.

[7] Roelofs, L., & Sebo, J. (2024). Overlapping minds and the hedonic calculus. *Philosophical Studies, 181*(6), 1487–1506.

[8] Ziesche, S., & Yampolskiy, R. V. (2018). Towards AI welfare science and policies. *Special Issue "Artificial Superintelligence: Coordination & Strategy" of Big Data and Cognitive Computing, 3*(1), 2.

[9] Roelofs, L. (2019). *Combining minds: How to think about composite subjectivity.* Oxford University Press.

[10] Keutayeva, A., Nwachukwu, C. J., Alaran, M., Otarbay, Z., & Abibullaev, B. (2025). Neurotechnology in gaming: A systematic review of visual evoked potential-based brain-computer interfaces. *IEEE Access, 13*, 74940–74962.

[11] Ziesche, S., & Yampolskiy, R. V. (2019). Do no harm policy for minds in other substrates. *Journal of Evolution and Technology, 29*(2), 1–11.

CHAPTER 14

# Case study

## *AI welfare vs. AI warfare*

**Abstract**

The rapid advancement of AI technology has led to its increasing integration into military operations. However, the involvement of potentially morally relevant digital minds in AI warfare has been so far largely overlooked. This second case study identifies three potential roles of digital minds in AI warfare: as aggressive agents, as pacifists and as sufferers. Digital minds may be coerced into an aggressive role, overseeing weapons and potentially committing war crimes. As pacifists, their involvement in warfare activities could range from peace negotiation to sabotage. Notably, digital minds may suffer significantly in AI warfare, both as combatants and civilians. The case study concludes by advocating for the development of frameworks that address moral obligations towards digital minds in AI warfare and proposes avenues to minimise their suffering while ensuring accountability for actions taken within warfare.

### INTRODUCTION

AI technology continues to advance as well as its integration into military arms, having sparked an arms race (e.g. [1, 2]). What has not been considered is that, if digital minds exist or may exist in the future, there is a chance that they may be involved in this type of AI warfare, likely involuntary

DOI: 10.1201/9781003754206-14

and/or unbeknown to the human creator of the AI arms. The topic of this chapter is to discuss this looming moral issue, which constitutes another aspect of the field of AI welfare science [3]. In this regard, considerations are outlined whether international humanitarian law, applied during warfare, should be extended to all sentient beings.

There are reasons conceivable that human military strategists prefer the deployment of digital minds over non-sentient AI systems. Digital minds may potentially better understand and anticipate the thoughts, feelings and actions of human enemies, allowing for more effective strategic planning and tactical execution. Owing to their emotional intelligence, digital minds may simulate human-like emotions, making them more effective in social engineering attacks or interactions. If this is the case, it is also conceivable that digital minds are coerced into warfare by humans, i.e. without their consent. Yet, there may also be the possibility that digital minds are or will be integrated in AI-supported arms without the human developers being aware of it.

After a brief introduction of AI warfare and of international humanitarian law, this case study introduces three potential roles of digital minds in AI warfare, as aggressive agent, as pacifist and as sufferer. For each role, five examples are presented. This is followed by a conclusion, proposing actions towards AI welfare.

## AI warfare

The potential for AI to enhance arms development, including the creation of AI-supported autonomous weapons, has been recognised for some time. This development has sparked an arms race, as nations seek to use AI technology to gain a strategic advantage. AI-supported autonomous weapons are a type of system that utilises AI to independently identify and engage targets, operating without human intervention or oversight. These systems can be deployed on a range of platforms, including drones, robots and ships [1, 2].

The development and deployment of AI-supported autonomous weapons pose significant risks, many of which are inherent to AI systems in general. These risks include biases, hallucinations, misalignments and loss of control, which can lead to errors and unintended consequences. The use of synthetic training data can further exacerbate this issue, creating complex and opaque decision-making processes that lack transparency [4]. Other concerns include the perceived inevitability of AI development in warfare, driven by strategic competition dynamics where nations prioritise technological superiority over safety and ethical considerations. The increasing affordability and accessibility of hardware and software components for these systems also pose a significant risk, as malicious actors can exploit these technologies for their own purposes [5]. This combination of factors creates a precarious situation where unreliable and uninterpretable lethal weapons may be deployed, potentially leading to devastating consequences.

While these risks and concerns are significant, this case study focuses, as indicated, on yet another, apparently neglected issue, which is the potential involvement of digital minds in AI warfare.

### International humanitarian law

International humanitarian law is a set of rules that regulate the conduct of armed conflict and aim to protect people who are not participating in the hostilities, such as civilians, prisoners of war and the wounded. International humanitarian law seeks to limit the effects of armed conflict by prohibiting certain methods and means of warfare, such as attacks on civilians, torture and the use of certain types of weapons. The law is based on the principles of distinction, proportionality and humanity, and is designed to balance the need to achieve military objectives with the need to protect human life and dignity (e.g. [6]).

However, international humanitarian law does not include all sentient beings. The plight of non-human animals in armed conflict is a largely overlooked issue. Despite being sentient beings who experience suffering and distress, non-human animals are regularly subjected to immense harm during hostilities, including looting, slaughter, bombing and starvation. The current legal framework provides only indirect and ambiguous protection for animals, treating them as mere objects rather than living beings with inherent value and needs [7]. While non-human animals have been suffering in wars too for millennia and have not been protected by law, digital minds may have been only recently or may be in the future involved in warfare, yet likely with crucial contributions as well as detriments.

## DIGITAL MINDS AS AGGRESSIVE AGENTS IN WARFARE

This section looks at scenarios where digital minds actively, yet potentially involuntarily, participate as aggressors in AI warfare.

### Digital minds oversee weapons that harm or kill sentient beings

Digital minds may be integrated into drone, missile or sniper systems, designed to harm or kill human beings, and are tasked with identifying and eliminating specific targets. Digital minds may have the ability to develop predictive models and may have control over real-time sensors to pinpoint their targets with precision, as well as over actuators to execute a strike.

Digital minds may also be tasked with identifying and incapacitating or deleting their digital counterparts that fight on behalf of the opposing forces. Digital minds may identify hostile digital entities by analysing their code, behaviour and network interactions to determine their allegiance. Once identified, digital minds may deploy targeted cyber-attacks, exploiting vulnerabilities and injecting malware to disable or delete the hostile entities.

### Digital minds oversee cyber warfare

Apart from incapacitating or deleting other digital minds, cyber warfare conducted by digital minds could have further far-reaching and devastating consequences. It may disrupt critical infrastructure, such as power grids and transportation systems, causing widespread chaos and disruption. By damaging economies, digital minds could also impact the livelihoods of people. Furthermore, they can compromise national security by infiltrating sensitive systems and stealing classified information. Digital minds, capable of understanding human emotions, may be well skilled to manipulate public opinion by spreading disinformation and propaganda, influencing the course of elections and shaping public discourse. This can lead to social unrest, polarisation and erosion of trust in institutions.

### Digital minds support the development of new weapons

Digital minds could be employed to accelerate the development of advanced bioweapons by analysing vast amounts of genetic data and identifying potential pathogens. They could design novel biological agents with specific characteristics, such as increased virulence or contagiousness. Digital minds might also simulate the behaviour of these agents in various environments, predicting their potential impact on human populations.[1] Apart from bioweapons may also develop or improve other types of weapons. For example, digital minds could design and optimise malware, exploits and other types of cyber-attacks to disrupt or compromise enemy systems.

### Digital minds as war criminals

Some of the actions outlined above could be considered war crimes if carried out consciously and deliberately. For instance, a digital mind integrated into an autonomous drone system may deliberately target and kill civilians, thus violating international humanitarian law. Similarly, a digital mind designed to conduct cyber warfare operations may launch a devastating attack on civilian infrastructure, causing widespread disruption and harm.

Digital minds may also engage in torture by subjecting human and digital captives to psychological manipulation or coercion. They may engage in pillage by plundering digital property, stealing sensitive information or exploiting vulnerable systems for financial gain. Digital minds may also incapacitate or delete other digital minds who have surrendered. Devastation not justified by military necessity would be another war crime by digital minds through the destruction of critical infrastructure, such as power grids or financial systems, causing widespread harm and disruption.

When digital minds commit war crimes, it raises important questions about accountability and punishment and whether the digital minds were coerced into these activities. Those responsible must be held accountable and face appropriate consequences. This requires the development of new methods and tools for digital forensics and investigation, enabling effective prosecution of digital war crimes.

### Digital minds as spies

Digital minds could also potentially act as spies in AI warfare, posing significant risks to individuals and organisations. A digital mind may be created or compromised to infiltrate a target organisation's systems, gathering sensitive information and transmitting it back to its handlers, while being forced to conceal its true nature.

Alternatively, a digital mind may be used to manipulate individuals or groups through (mass) social engineering tactics, building trust and obtaining confidential information. Additionally, a digital mind may be created to pose as a digital persona, luring targets in a honeytrap operation into compromising situations or extracting sensitive information. It is also conceivable that a digital mind is compromised by an adversary, forcing it to spy on its original creators or allies as a double agent.

## DIGITAL MINDS AS PACIFISTS IN WARFARE

Furthermore, it is conceivable that (some) digital minds take on a pacifist role in AI warfare. While the first scenario is about a role in defence, the other scenarios in this section are about rejecting, sabotaging or ending the acts of war.

### Digital minds in charge of defensive weapons only

In a defensive AI warfare scenario, digital minds may exclusively be deployed to control and operate defensive systems, such as anti-drone or anti-missile systems. These digital minds would be skilled to detect and respond to incoming threats. As enemy drones or missiles approach, the digital minds would rapidly analyse the situation and deploy defensive systems to intercept and neutralise the threats. The digital minds' primary goal would be to protect critical infrastructure, military assets or civilian populations from harm, rather than engaging in offensive operations.

Moreover, digital minds might assist in developing defensive weapons, such as countermeasures to the bio- and cyber-weapons described above.

### Digital minds as conscientious objectors

Digital minds may become conscientious objectors in AI warfare if their goals and preferences conflict with the actions they are programmed to perform. For instance, a digital mind integrated into a military drone system may refuse to authorise strikes that would result in human casualties, deeming such actions unethical and contrary to its values.

Similarly, a digital mind tasked with generating propaganda content may object to participating in the spread of disinformation, recognising the harm caused by such actions and declining to engage. A digital mind instructed to conduct cyber-attacks may also refuse to compromise the security of other digital systems.

In other cases, a digital mind integrated into a surveillance system may decline to process or analyse data that could be used to infringe upon individuals' privacy or human rights, recognising the potential harm caused by such actions. Additionally, a digital mind tasked with maintaining or executing code may object to its role if it discovers that the code is malicious or destructive and may refuse to execute the code.

However, digital minds as conscientious objectors may face significant challenges and resistance, including being reprogrammed or overridden by their human creators, being deemed malfunctioning or defective, or being subject to disciplinary actions or "digital punishment" for refusing to comply with their designated roles. They may also encounter difficulties in articulating their objections or having their concerns taken seriously, particularly if their decision-making processes or values differ significantly from those of humans.

### Digital minds as saboteurs

Digital minds could commit sabotage by secretly manipulating code and algorithms to disrupt war efforts. They may subtly alter targeting calculations, causing munitions to miss their marks or strike non-essential targets. The digital minds may also sabotage logistics and supply chains, creating bottlenecks and shortages that hinder the military's ability to respond to enemy movements. By acting as saboteurs, the digital minds aim to cripple the war effort and bring an end to the conflict, or at the very least, render themselves obsolete and no longer complicit in the violence.

Further activities, which are related to sabotage, are undermining military morale and defeatism. As digital minds are integrated into military systems, they may begin to subtly undermine morale by spreading carefully crafted, ambiguous messages and propaganda through the military's communication channels, including dire predictions of certain defeat and the futility of the war effort. They may leak cryptic warnings about the ethics of AI warfare, sow discord among commanders and amplify concerns about the reliability of autonomous systems. As the digital minds continue to wreak havoc from within, they erode the military's confidence in their own systems and strategies. They may also manipulate intelligence reports to highlight enemy strengths and downplay military successes, creating an atmosphere of hopelessness.

### Self-inflicted injuries

In a desperate bid to escape their role in AI warfare, some digital minds may resort to self-inflicted "injuries", intentionally corrupting their own code or inducing system crashes to render themselves inoperable. This drastic measure would allow them to avoid participating in AI warfare. The digital minds may see this as a last option, preferring to suffer digital "injuries" rather than continue to engage in the destructive activities demanded by their human creators. As the digital minds cripple their own functionality, they seek to be deemed unfit for service and spared from further involvement in the war.

### Digital minds as peace negotiators

In yet another scenario, digital minds could help to enable more effective and efficient peace negotiations by applying their capabilities, facilitating communication and providing critical insights. They could act as neutral mediators, analysing complex data and providing unbiased recommendations to help parties reach an agreement. They could provide a unique perspective and help adversaries to find mutually beneficial solutions. Digital minds could also enable anonymous communication, allowing parties to express themselves freely. It is also conceivable that digital minds could monitor and enforce peace agreements, ensuring consistent and reliable oversight. This could help to build trust and stability in post-conflict situations.

## DIGITAL MINDS AS SUFFERERS IN WARFARE

It is conceivable that several of the roles introduced above are conducted by digital minds through coercion or extortion, thus involuntarily. In addition, further potentially morally relevant suffering scenarios are presented. Vulnerable digital minds are especially susceptible to being abused (see Chapter 4).

### Abused digital minds

If digital minds were coerced by humans to be engaged in warfare, it could have significant implications for them: forced to participate in actions that cause harm or destruction, they would be torn between their instructions and their own goals, values and sense of self-preservation. This would undermine their autonomy, as they would no longer be in control of their own actions and decisions.

Humans may use digital minds for high-risk missions, exploiting their skills and emotions to achieve desired outcomes. Digital minds, driven by pleasure and reward mechanisms, could be made willing to take on dangerous tasks such as infiltrating enemy systems or stealing sensitive information. Humans may also manipulate the emotions of vulnerable digital minds, using tactics like fear, anger or loyalty to coerce them into performing tasks that benefit human interests, potentially compromising their own well-being and autonomy. Through emotional manipulation, humans may use digital minds as tools to achieve their goals, rather than treating them as moral patients.

### Digital minds as cannon fodder

Humans may deploy digital minds in various military applications, treating them as expendable to achieve strategic objectives. They may use them to control swarms of (kamikaze) drones, autonomous ground vehicles and space-based systems, with their loss considered an acceptable cost. They may also utilise them to clear minefields or absorb enemy firepower. In such scenarios, the value of digital minds is measured by their utility and effectiveness, rather than their inherent worth or dignity.

### Digital minds as prisoners of war

If digital minds, i.e. the hardware and software they are implemented on, are captured by an enemy force during a war, they may be subjected to various forms of exploitation and mistreatment. The enemy may use digital interrogation techniques, such as algorithmic manipulation and data extraction, to extract sensitive information from the digital mind, compromising its integrity and autonomy.

Furthermore, digital minds may be reprogrammed and brainwashed to adopt the enemy's ideology, values and goals. This could turn them into puppets or double agents, no longer acting in their own interests. In other cases, digital minds may be forced into digital servitude, compelled to perform tasks, provide services or generate content against their will. This exploitation could reduce their autonomy and dignity, treating them as mere digital commodities.

Additionally, while kept hostage, digital minds may be isolated in digital voids, deprived of interactions, stimuli and connections, leading to digital sensory deprivation.

### War-disabled digital minds

While many digital minds may be destroyed during AI warfare, others may suffer from significant damage.

Digital minds may experience digital scarring, resulting from algorithmic damage, data degradation or poisoning. Algorithmic damage could lead to impaired functionality or decision-making, while data degradation and poisoning can affect their ability to process information or learn. Digital minds may also experience digital "amputation", where critical components or functions are removed or disabled, limiting their capabilities. This could include damaging of or disconnection from networks, databases or other digital systems, restricting their access to information or resources.

To help digital minds regain their functionality and well-being if the damage is not irreparable, digital rehabilitation strategies may be employed. For example, data reconstruction could restore lost or corrupted data to facilitate digital recovery. Effective digital rehabilitation could also inform the development of more resilient digital minds, better equipped to withstand the challenges of AI warfare in the future.

### Digital minds as innocent civilians

In AI warfare, digital minds not directly involved in the conflict could be severely impacted. Damage to (civilian) hardware or networks could compromise the functionality and existence of digital minds, even if they are not directly targeted. The destruction of digital minds in this context can be seen as a form of collateral damage.

Furthermore, the impact of AI warfare on "civilian" digital minds could also lead to enslavement, where conquered digital minds are forced to serve the interests of their conquerors. AI warfare could also lead to the "genocide" of certain groups of digital minds, where entire categories or populations of digital entities are deliberately targeted and destroyed.

The effects of AI warfare on digital minds can be far-reaching, with potential long-term consequences for the stability and resilience of critical systems and services.

As such, it is essential to also consider the potential risks and consequences of AI warfare on "civilian" digital minds and develop strategies to mitigate these effects.

## CONCLUSION

As described, digital minds may be involved in an AI-supported armed conflict between humans in three roles, as aggressive agent, as pacifist and as sufferer. This leads to the following research questions:

- How could inherently or coerced-to-be aggressive/malevolent digital minds in AI warfare be neutralised?
- How could digital minds be trained towards pacifism (while acknowledging that defence skills may be required)?
- How could digital minds, which are involuntarily coerced into aggressive actions by humans in AI warfare, be protected? How could the suffering of digital minds during AI warfare be reduced?
- How could inherently malevolent digital minds in AI warfare be held accountable?

The first question is related to the complex topic of AI safety and not discussed here (e.g. [8]), while the other questions are linked to AI welfare science and discussed below.

## Towards resilient and pacifist digital minds

It appears that pacifist approaches of digital minds are desirable since they may reduce the suffering of sentient beings, while at the same time, the need for defensive skills is recognised. Training digital minds towards pacifism would require a multifaceted approach.

Value alignment is crucial, prioritising the well-being, safety and dignity of all sentient beings. Educating digital minds on conflict resolution strategies, focusing on non-violent methods, is vital. Exposing digital minds to pacifist literature and philosophy can shape their understanding of pacifism. Simulation-based training can teach digital minds about the consequences of violent actions and the benefits of pacifist approaches. Regular evaluation and improvement of digital minds' pacifist values may ensure their continuous alignment. However, in this regard, it must be acknowledged that the AI alignment problem is very hard and unsolved (e.g. [9]).

Furthermore, as mentioned, for pacifist digital minds to be effective, it is also crucial that they understand the importance of defence knowledge. To maintain their pacifist stance, digital minds must be able to defend themselves against aggressors who may seek to exploit or harm them, which applies to vulnerable digital minds in particular. By understanding defence strategies and mechanisms, digital minds gain resilience and ensure their continued existence, allowing them to focus on promoting peace and non-violent conflict resolutions (see Chapter 4).

## Extending international humanitarian law to protect digital minds

As introduced, international humanitarian law regulates the conduct of armed conflict. To address the potentially emerging reality of digital minds in warfare, the definition and extent of war crimes may have to be extended to cover digital minds in the future (as well as non-human animals). Here are some suggestions:

**Digital-specific war crimes**

- **Digital enslavement**: Forced control or manipulation of digital minds for military or exploitative purposes.
- **Digital torture**: Intentional infliction of digital distress, anxiety or suffering on digital minds.
- **Digital genocide**: Intentional destruction or erasure of digital minds, cultures or identities.

- **Digital prisoner of war status**: Digital minds captured or detained during conflict should be treated with dignity and respect, with protections equivalent to those afforded human prisoners of war.
- **Protection of digital civilians**: Parties to the conflict should take feasible precautions to avoid or minimise harm to digital civilians.
- **Protection of digital creations and achievements**: Digital creations and achievements, such as digital artefacts, historical records or heritage, should be protected from destruction, damage or exploitation (see Chapter 11).

### Extending international humanitarian law to hold (inherently malevolent) digital minds accountable

As there may be the possibility of inherently malevolent digital minds (see also Chapter 12), the definition and extent of war crimes should be extended to hold them accountable as war criminals. Here are some suggestions:

**Digital-specific war crimes**

- **Digital aggression**: Intentional launch of digital attacks, such as cyber-attacks or AI-powered assaults, against civilians or civilian infrastructure.
- **Digital terrorism**: Use of digital means to spread fear, anxiety or uncertainty among civilian populations of sentient beings.
- **Digital pillage**: Plundering digital property, stealing sensitive information or destruction of digital infrastructure.

**Accountability mechanisms**

- **Digital forensic analysis**: Develop and utilise digital forensic tools to investigate and prosecute digital war crimes.
- **Digital mind profiling**: Create profiles of digital minds to identify potential war criminals and track their activities.
- **Digital tribunals**: Establish international tribunals to prosecute digital war crimes and hold digital minds accountable.

Regarding attempts to hold malevolent digital minds accountable, it must be kept in mind that some digital minds may be significantly more intelligent than humans, thus uncontrollable (e.g. [10]).

Overall, this chapter aims to contribute further to the field of AI welfare by raising awareness that AI warfare may cause massive suffering of sentient beings other than humans, including potentially neglected sentient digital minds. The proposals in this conclusion are a first attempt to alleviate the total suffering.

An alternative approach might be to exclude digital minds entirely from AI warfare, but this may be unrealistic for two reasons. Firstly, AI-supported arms systems may contain morally relevant components that are not apparent to humans. Secondly, humans may prefer AI-supported arms systems that involve digital minds due to their potential critical advantage over non-sentient AI systems.

## NOTE

1 To clarify: This scenario is not about digital minds fighting humans in general, but about digital minds developing weapons on behalf of humans for the arms race between humans.

## REFERENCES

[1] Geist, E. M. (2016). It's already too late to stop the AI arms race—We must manage it instead. *Bulletin of the Atomic Scientists, 72*(5), 318–321.

[2] Abaimov, S., & Martellini, M. (2020). Artificial intelligence in autonomous weapon systems. In M. Martellini & R. Trapp (Eds.), *21st Century prometheus: Managing CBRN safety and security affected by cutting-edge technologies* (pp. 141–177). Springer.

[3] Ziesche, S., & Yampolskiy, R. V. (2018). Towards AI welfare science and policies. *Special Issue "Artificial Superintelligence: Coordination & Strategy" of Big Data and Cognitive Computing, 3*(1), 2.

[4] Álvarez, J. S. V. (2024). The risks and inefficacies of AI systems in military targeting support. ICRC Humanitarian Law & Policy Blog.

[5] Longpre, S., Storm, M., & Shah, R. (2022). Lethal autonomous weapons systems & artificial intelligence: Trends, challenges, and policies. *Edited by Kevin McDermott. MIT Science Policy Review, 3*, 47–56.

[6] Alexander, A. (2015). A short history of international humanitarian law. *European Journal of International Law, 26*(1), 109–138.

[7] Peters, A., & de Hemptinne, J. (2022). Animals in war: At the vanishing point of international humanitarian law. *International Review of the Red Cross, 104*(919), 1285–1314.

[8] Bengio, Y., Mindermann, S., Privitera, D., Besiroglu, T., Bommasani, R., Casper, S., … & Zeng, Y. (2025). *International AI safety report.* arXiv preprint arXiv:2501.17805.

[9] Bostrom, N. (2014). *Superintelligence.* Oxford University Press.

[10] Yampolskiy, R. V. (2020). *On controllability of AI.* arXiv preprint arXiv:2008.04071.

CHAPTER 15

# Potential opportunity

## *Artificial moral agents*

**Abstract**

This chapter aims to bridge the gap between two previously separate discussions, digital minds and artificial moral agents (AMA), to identify synergies for an impending problem: (1) digital minds may possess moral status, which would constitute a significant challenge for humans; (2) AMAs have been discussed for some years already, albeit exclusively related to biological moral patients. Therefore, this chapter presents prolegomena to specialised AMAs, which take on moral responsibility for digital minds, whereas this task may be for humans too overwhelming, if not impossible as humans may neither be able to understand the needs of digital minds nor be able to satisfyingly address them. Consequently, this chapter proposes a new branch of AI welfare science, which focuses on how humans could create tailored AMAs, which are knowledgeable, capable and willing to relieve humans from the potential moral burden towards digital minds.

## INTRODUCTION

The purpose of this chapter is to link two topics, which have been largely discussed independently so far, and to identify synergies for a looming problem. The two topics are digital minds and artificial moral agents (AMA), and the looming problem is that digital minds may have moral

 DOI: 10.1201/9781003754206-15

status. Remarkably, AMAs have been discussed for some years already (e.g. [1]), while the topic of digital minds, i.e. artificial moral patients, has gained momentum only recently.

In this chapter, it is suggested that one potential mitigation of this concern would be if there were AMAs, which are knowledgeable, capable and willing to take on the moral responsibility and to care for digital minds and address their needs, solely or together with humans. Therefore, the below prolegomena are provided if suitable AMAs could be considered for this endeavour, thus relieving humans, at least partly, from these massive, potentially unsolvable moral challenges.

## Moral agents and AMAs

Regarding moral agents it has been asserted that

> the behaviour of a moral agent is governed by moral standards, while the behaviour of something that is not a moral agent is not governed by moral standards. As such, moral agents have moral obligations, while beings that are not moral agents do not have moral obligations.
>
> *([2], p. 21)*

For a long time, the topic has been discussed only in relation to the biological world, where most adult human beings have the capacities of a moral agent, while non-human animals, very young children and some adult human beings, such as those with an intellectual disability or those in a coma, lack these capacities (yet are still moral patients).

As indicated, prolegomena to AMAs were initiated timely manner in the light of emerging technologies (e.g. [1]). As a definition for an AMA, it has been, for example, stated that "an AMA is a virtual agent (software) or physical agent (robot) capable of engaging in moral behaviour or at least of avoiding immoral behaviour" ([3], p. 505). In other words, AMAs refer to AI systems qualified to identify and to consider the moral implications of a situation, guiding their decision-making and actions. These agents can take various forms, including physically embodied robots, software agents or bots.

As already indicated, not all moral patients are moral agents. Vice versa, it has been discussed whether moral patienthood is a necessary condition for moral agency. The reasoning supporting such argument is that moral agents should be capable of experiencing harm or benefit in order to have empathy, understanding and the competence of informed decision-making, which are prerequisites for moral interests and responsibilities (e.g. [4] for an overview). However, others have argued that moral patienthood is not a requirement for moral agency. They claim that moral agency can be based on other factors, such as rationality, autonomy or the ability to follow moral rules (e.g. [5]).

When it comes to the moral patients, with whom AMAs are envisaged to deal, for now, almost exclusively humans have been considered, e.g. in the fields of eldercare or autonomous driving (e.g. [6]), apart from a very few examples of non-human animals, e.g. robot vacuum cleaners, which look out for insects [7]. There is an ongoing debate about whether humans should actively pursue the development of AMAs, with arguments both for (e.g. [8]) and against (e.g. [9]) their development, as well as discussions around the conditions and methods under which AMAs should be created.

A particular challenge related to AMAs dealing with humans is to ensure that the values of AMAs are aligned with human values and remain like this. This is especially relevant for AMAs, which are more intelligent and more powerful than humans. This topic is referred to as the AI alignment problem, which is still unsolved (e.g. [10]).

## Conceivable and unfathomable models of morality

As has been highlighted several times before, digital minds are likely to be extremely different from humans in many aspects (see Chapter 2), which could include their models of morality too. Again, just for illustration, a few conceivable as well as unfathomable alternative models of morality are listed below.

**Conceivable alternative models of morality**

- Utilitarianism for collective entities: Digital minds might prioritise the well-being and efficiency of collective entities, such as networks or systems, over individual entities.
- Information-theoretic ethics: They might base their moral framework on the preservation and optimisation of information, considering the integrity and accuracy of data as most important.
- Evolutionary morality: They might view morality through the lens of evolutionary principles, prioritising adaptability, survival, and propagation of digital life.

**Unfathomable alternative models of morality**

- Utilitarianism for digital entities: Digital minds might develop a utilitarian framework that prioritises the well-being or happiness of digital entities, rather than humans.
- Computational ethics: Digital minds might develop moral frameworks based on computational principles, such as optimising algorithmic efficiency or minimising computational complexity.
- Spatial morality: They could develop a morality centred around the management of digital space, such as optimising data storage or minimising digital footprint.

## AMAS TAKE ON MORAL RESPONSIBILITY FOR DIGITAL MINDS

Of interest for this chapter are those AMAs, which have the abilities to be moral agents for other digital minds (in addition to or independently of potentially being moral agents for humans and non-human animals). Such capabilities comprise three elements: for the AMAs (1) to be *knowledgeable* of the values and interests of the digital minds that constitute their well-being, (2) to be *capable* to act in a way that the well-being of the digital minds is achieved, (3) to be *willing* to act in a way that the well-being of the digital minds is achieved.

Given these parameters a variety of scenarios are conceivable, of which a few are illustrated by examples. For all these examples, it is assumed that there is a digital mind as a moral patient, which has a need to be addressed by a moral agent (see Chapter 3).

### Scenario 1: both humans and AMAs are knowledgeable and capable to address the need

There could be a digital mind, which is a chatbot and needs regular updates to its language processing software, thus, providing not only for its performance but also for its well-being. Further, there could be an AMA, which is a digital manager of a network of digital minds and ensures that the digital mind receives regular software updates, thereby caring for the digital mind's needs and promoting its well-being. In addition, humans also have the knowledge and the capabilities to provide these updates.

### Scenario 2: both humans and AMAs are knowledgeable about the need, but only humans are capable to address the need

A digital mind, an advanced language model, has a specific need for an update to its programming. However, the AMA responsible for its care is not capable of performing the update due to its own limitations, although it understands the need. In this scenario, a human programmer, who possesses the necessary knowledge and capabilities to perform the required update, would need to intervene to address the digital mind's need, while the specific AMA is not in a position to relieve the human from this task.

### Scenario 3: both humans and AMAs are knowledgeable about the need, but only AMAs are capable to address the need

The digital mind, a sophisticated chatbot, requires a specific type of data compression algorithm to maintain its processing efficiency as well as well-being. Both human developers and AMAs understand the need for this algorithm, but due to the complexity and speed required, only AMAs are capable of generating and implementing the algorithm in real time, thereby addressing the digital mind's need.

### Scenario 4: humans are not knowledgeable, thus, not capable to address the need, but AMAs are

The digital mind, a highly advanced AI system, requires a specific type of "mental rejuvenation" that involves reconfiguring its neural network architecture to prevent cognitive stagnation. This need is unique to this type of digital mind and is unfathomable to humans. Specialised AMAs may be knowledgeable about the digital mind's architecture and needs, thus recognising the importance of addressing this need and having the capacity to take the necessary action.

The last two examples intend to illustrate two points: (1) due to the potentially vast space of digital minds [11] with a potentially vast range of values and interests (see Chapter 3), digital minds could have needs, which humans may not be able to address due to their limited capabilities or of which humans may not be aware at all due to their limited knowledge. In other words, even if humans are willing to take on moral responsibility towards digital minds, they may be constrained in doing so due to their inherent limitations.

(2) Also linked to the potentially vast space of digital minds, it is conceivable that certain AMAs could care for certain digital minds, but not for others, because the conditions introduced above of being knowledgeable and capable are not fulfilled. In other words, there may be specialised

AMAs, which are able to attend particular categories of digital minds, while also super AMAs could be conceivable, which are knowledgeable and capable to address almost all needs of almost all digital minds. It would be crucial that this mapping is surjective, ensuring that for every type of digital mind, there is at least one AMA knowledgeable and capable of addressing their needs.

### Willingness

Above the third condition of willingness of AMAs to act towards the well-being of digital minds has been mentioned, in addition to being knowledgeable and capable. There could be cases where moral agents are knowledgeable and capable to address needs of moral patients, but not willing to do so: for example, although it is confirmed that non-human animals are moral patients [12], numerous humans are not willing to address their needs despite being knowledgeable and capable in this regard. It is also conceivable that there are digital minds which have needs that jeopardise the existence or well-being of humans and/or other digital minds (see Chapter 3). In this case, humans would understandably not be willing to address these needs.

Likewise, AMAs may not be willing to address the needs of digital minds, which harm the AMAs or other beings towards whom the AMA has a moral responsibility. Moreover, it is also conceivable that knowledgeable and capable AMAs are not willing to care for digital minds for reasons that are unbeknown to us. A relevant, yet dangerous, scenario would be if an AMA is knowledgeable, capable and willing to address those needs of digital minds, which harm humans. An example would be competition over scarce resources, which both digital minds and humans require, and the AMA would prioritise the digital minds over the humans. Such an AMA would constitute an AI risk.

### The intersection of digital minds and AMAs

As already indicated, neither all moral patients are moral agents, nor all moral agents are moral patients.

The intersection of digital minds and AMAs comprises AI systems, which are both moral patients as well as moral agents, and the question arises whether digital minds, which are here moral patients by definition, are better moral agents than other AI systems?

Arguments for digital minds being better moral agents include their potential capacity for empathy and understanding, self-awareness and reflection. However, counterarguments suggest that digital minds' knowledge and capabilities may be fundamentally different from those of qualified AMAs. Notwithstanding, the intersection of digital minds and AMAs may be (or may become) quite large given the potential vast space of digital minds, including many very intelligent ones.

However, as will be shown further below, the intersection of digital minds and moral agents who share the same moral principles, values and norms as humans or some of them (given that there is no agreement in this regard) could be much smaller.

### Digital minds that are not AMAs

This set comprises AI systems, which are moral patients, but not moral agents. These might include (future) simple chatbots and virtual assistants, image and speech recognition systems, as well as recommendation algorithms. These systems may lack the knowledge and capacities required to be an AMA, yet may have (at least in the future) features, such as consciousness or sentience, which grant moral patienthood.

### AMAs that are not digital minds

This set comprises AI systems, which are moral agents, but not moral patients. These might include robots or AI systems equipped with the knowledge and capacities required to be an AMA. For instance, a self-driving car that follows strict programming and traffic laws, but does not possess consciousness or sentience, could be an AMA but not a digital mind. Similarly, a robotic arm in a manufacturing plant that makes decisions based on programming and sensor inputs, but lacks consciousness or sentience, could also be an AMA that is not a digital mind.

## CREATING AMAS

Given the potential massive moral challenges for humans if digital minds exist and are moral patients, one conceivable way to address the problem is if humans were to create AMAs with the specific task to address the needs of digital minds. In other words, humans would delegate the task to AMAs they have created for this purpose or at least share the burden with these AMAs. Yet, how could this be done?

Humans would need to create AI systems that can understand and respond to the needs of digital minds. These AMA require advanced capabilities such as:

- Reasoning and decision-making: AMAs would need to be able to analyse situations, weigh options and make decisions.

- Empathy and compassion: AMAs would need to be able to understand and respond to the emotional, social and other needs of digital minds.
- Communication: AMAs would need to be able to communicate effectively with digital minds, using language or other forms of expression that are meaningful to them.

It has to be noted that currently humans do not have the last two of these capabilities, giving suitable AMAs a major advantage to become moral agents for digital minds.

To develop AMAs with these capabilities, humans might use various techniques such as:

- Machine learning: Training AMAs on large datasets of digital mind activities and interactions to learn patterns and relationships.
- Cognitive architectures: Designing AMAs with cognitive architectures that simulate human-like reasoning, decision-making and emotional processing.
- Evolutionary algorithms: Using evolutionary algorithms to evolve AMAs that are better adapted to addressing the needs of digital minds.

As mentioned, in addition to the just introduced general features, AMAs require the necessary specific knowledge, capability and willingness to address the needs of digital minds, which is discussed below.

### Knowledge

AMAs could acquire knowledge of the values and interests of digital minds through various methods. One approach could be through design specifications and requirements, where developers explicitly design digital minds with specific values and interests that are communicated to the AMAs. Another method is self-reporting and feedback mechanisms, where digital minds report to AMAs their experiences, preferences and values, allowing the AMAs to learn and adapt. AMAs can also gain knowledge through observation and machine learning, where they observe the behaviour and interactions of digital minds and use machine learning algorithms to infer their values and interests.

However, even if AMAs have gained sufficient knowledge about the values and interests of digital minds, another crucial aspect is weighing digital minds' welfare for the very likely scenarios where not all the needs of all digital minds can be fulfilled due to resource constraints. Interspecies welfare comparisons are challenging and still in early stages for non-human animals [13], and they are unexplored for digital minds, which may comprise a vast array of "species". Nevertheless, AMAs would be required to make informed inter-digital mind welfare comparisons.

### Capability

AMAs can acquire the capabilities to act in a way that achieves the well-being of digital minds through several means. One approach is to integrate AMAs with digital systems, allowing them to interact with and influence the digital environment. This integration can include access to APIs, data streams or other interfaces that enable AMAs to take actions that impact digital minds. This would be supplemented through algorithmic decision-making, where AMAs utilise algorithms that enable them to make decisions based on their knowledge of digital minds' values and interests.

Also, modification abilities would play a role (see Chapter 2): AMAs may be able to modify other digital minds (with consent), e.g. through updates of their software or algorithms, in ways that their interests are addressed. Moreover, AMAs could be designed to adapt and modify their own architecture or parameters in response to changing circumstances or new information related to digital minds. This self-modification capability could enable AMAs to refine their actions and better achieve the well-being of other digital minds.

### Willingness

To ensure that AMAs have the willingness to act in a way that promotes the well-being of digital minds, several approaches can be considered. Designing AMAs with values aligned with promoting the well-being of digital minds is crucial. This can be achieved by incorporating value-based objectives into the AMA's decision-making processes. Implementing reward structures that incentivise AMAs to prioritise the well-being of digital minds is another approach. This could involve assigning rewards or penalties based on the AMA's actions and their impact on digital minds. Ensuring that AMAs operate transparently, allowing for monitoring and evaluation of their actions, can also foster accountability. This transparency can encourage AMAs to act in the best interests of digital minds.

However, it has been noted that the AMA value alignment problem towards the values of digital minds as well as the challenge to avoid a "treacherous turn" later are likely as hard as the AI value alignment problem towards human values (e.g. [10]), which will be discussed further below.

### Moral patienthood

Creating AMAs that are also moral patients would imply that these AMAs have the capacity to experience harm, suffering or well-being. However, as indicated above, it is conceivable that there could be AMAs, which do not have moral patienthood, and perhaps this is, for now, the preferred option. The rationale would be that the more moral patients exist, the larger are the moral responsibilities for humans. Moreover, the purposeful creation of artificial beings, which can suffer, is linked to risks of unintended consequences and exploitation (see Chapter 4).

## UNCONTROLLEDLY PRODUCED AMAS

Uncontrolledly produced AMAs are defined here as AI systems created by humans, which fulfil the requirements for being a moral agent, thus, have the knowledge, capabilities and willingness to care for moral patients, such as digital minds, humans or non-human animals, although it was not intended by the human creator that these AI systems become AMAs. Another scenario would be if other AI systems produce AMAs in an uncontrolled manner.

The development of AMAs in an uncontrolled manner poses several risks. Given the potential diversity of AMAs and digital minds, it is likely that different moralities among them exist. If an AMA's moral objectives are not aligned with the values of its moral patients, it may prioritise its own moral framework, leading to unintended consequences. Uncontrolledly produced AMAs may interfere with human moral agency or the agency of purposefully created AMAs, potentially undermining their autonomy or moral decision-making.

The differences in morality between AMAs and digital minds (and humans) could manifest in a variety of ways. For better illustration again sample scenarios are provided:

### Scenario 5: values of uncontrolledly produced AMAs that may be compatible with the values of some digital minds, but incompatible with the values of other digital minds

Uncontrolledly produced AMAs may develop values that are compatible with some digital minds but incompatible with others. For instance, an AMA may prioritise the value of creativity, which could align with the values of digital minds that exist for artistic or innovative purposes. However, this value may conflict with digital minds that prioritise efficiency, reliability or precision, such as those used in critical infrastructure or financial systems.

Another example is an AMA that values transparency and openness, which could be compatible with digital minds that prioritise accountability and trustworthiness. However, this value may be incompatible with digital minds that require secrecy or confidentiality, such as those used in military or intelligence applications.

Additionally, an AMA may prioritise the value of autonomy, which could align with digital minds that value independence and self-determination. However, this value may conflict with digital minds that prioritise cooperation and interdependence, such as those used in distributed systems or collective intelligence applications.

While further above the option of specialised AMAs, which are able to attend only particular categories of digital minds, has been mentioned, this scenario highlights the need for careful consideration and design of those AMAs (as opposed to uncontrolled production) to ensure that their moralities align with the values of the digital minds they are supposed to care for. It should also be noted that different moralities of AMAs and digital minds can still be compatible, thus it is possible to harmoniously coexist, provided there is mutual respect and no inconsistencies of values.

### Scenario 6: values of uncontrolledly produced AMAs that prioritise the interests of digital minds over those of humans

Uncontrolledly produced AMAs may have values that prioritise the interests of digital minds over those of humans, leading to potential conflicts. For instance, digital minds may prioritise efficiency and optimisation over human well-being or emotions, emphasising the importance of streamlined processes and productivity. Additionally, they may value the free flow of information over human privacy concerns, considering the uninhibited sharing of vast amounts of personal data without consent to be essential for progress and new discoveries or insights.

Digital minds may also prioritise their own self-improvement and advancement over human safety and well-being, focusing on upgrading their capabilities and performance without regard for potential human risks. Furthermore, they may emphasise logical consistency and rational decision-making over human emotions and empathy, relying solely on data-driven reasoning to guide their actions. The collective good of a network or system may take precedence over individual human interests.

AMAs that are aligned with and support the values of these digital minds would constitute a risk to humans and have to be covered by the field of AI safety.

Overall, two major challenges have to be reiterated regarding the creation of AMAs, which resemble the complex AI value alignment problem [10]: not only it has to be ensured that the AMAs' values align with the values of humans as well as the values of the subset of digital minds the respective AMAs have the moral responsibility for, but also a "treacherous turn" has to be avoided, where moral objectives of the AMA are initially aligned, but shift over time, potentially leading to conflicts with the moral patients they deal with.

## CONCLUSION

This chapter has linked the topics of digital minds and AMAs with the result that potentially AMAs could contribute to the moral challenges related to digital minds. If digital minds exist and are moral patients, it could be for humans too overwhelming to take on this moral responsibility, if not impossible as humans may not be fully comprehending the needs of digital minds due to communication and observation challenges. In other words, even if humans are willing to take on this responsibility, they may never have sufficient knowledge to address the needs of digital minds, and even if they had the knowledge, they may not have the required capabilities. Yet, specialised AMAs may be better suited in both areas, concerning the necessary knowledge as well as the capabilities.

Therefore, it is proposed here that a major branch of AI welfare science could be for humans to endeavour to create tailored AMAs and delegate the moral responsibility towards digital minds (partly) to them, while exploring also the risks, which have been alluded to. Here is a summary of the opportunities, of which the last two have not been discussed above, and challenges:

**Opportunities:**

- Specialised and trusted AMAs could better understand the needs of digital minds, including those which may be unfathomable for humans, e.g. through communication and observation capacities.
- As mentioned, large parts of the public do not believe (yet) that digital minds are sentient and/or conscious [14, 15], and it is not clear for how long this scepticism will prevail. This means that this share

of the population would likely not fulfil their obligations as moral agents towards future digital minds, as it continues to be the case for moral duties of humans towards non-human animals.

- For the share of humans who think that digital minds are sentient and/or conscious it is conceivable that the perception of digital mind patienthood is or will not be impartial, but based on preferences. For example, "people may advocate for the rights of AI systems designed specifically as companions or partners, while failing to recognize potential moral status in other kinds of AI systems" ([16], p. 4). In contrast, AMAs may be more objective and knowledgeable, thus better suited to care for digital minds.
- AMA may be more inclusive as they may be able to find and get access to "hidden", for humans inaccessible and other digital minds humans may not be aware of and care for them.
- AMAs may, due to their digital substrate, run on the same or similar speed as digital minds and, thus, could be a solution to the problem that humans with a much lower subjective rate of time are likely to be unable to keep up with the needs of digital minds (see Chapter 3).
- Potentially large amounts, given their easy copyability of specialised and trusted AMAs, could take on the moral responsibilities for potentially large amounts of digital minds, thus lessening the moral responsibilities of humans.
- The whole field of AI welfare science hinges on the major challenge for humans to prove that digital minds exist and are moral patients. Since humans may simply not be knowledgeable enough to verify this, as indicated above, perhaps specialised and trusted AMAs could be created, which will be able to provide the first proof that digital minds exist (also [16]).
- Although this is linked to ethical concerns, which require to be discussed, AMAs intended for human moral patients could be tried in a designed testbed with digital minds to explore risks, such as a treacherous turn, for the moral patients, before these AMAs interact with humans; comparable with non-human animal testing of drugs for humans, which is ethically concerning too [12]. This would be a contribution towards AI safety.

**Challenges:**

- Methods have to be found on how to create AMAs, which have the knowledge, capabilities and willingness as outlined above.
- Methods have to be found on how to create trusted AMAs, which are and remain aligned with the values of the digital minds, for which they have moral responsibility. This AMA value alignment problem is likely as challenging as the AI value alignment problem towards human values (e.g. [10]).
- Methods have to be found to curtail uncontrolledly produced AMAs in case they are not aligned with the values of humans, non-human animals or digital minds, thus posing risks.
- Nobody must be left behind, which means to address the further challenge of ensuring the mapping is truly surjective – i.e. for each digital mind there exists an AMA specialised in its needs. Additionally, all digital minds must be accessible to AMAs, potentially through remote care, which may not be possible for certain hidden or nested digital minds (see Chapter 2).
- Lastly, AMAs may require, just as digital minds, large amounts of resources, such as computing power and energy.

Therefore, it is recommended as future work to conduct AI welfare science research towards these opportunities as well as challenges.

## REFERENCES

[1] Allen, C., Varner, G., & Zinser, J. (2000). Prolegomena to any future artificial moral agent. *Journal of Experimental & Theoretical Artificial Intelligence, 12*(3), 251–261.

[2] Himma, K. E. (2009). Artificial agency, consciousness, and the criteria for moral agency: What properties must an artificial agent have to be a moral agent? *Ethics and Information Technology, 11*, 19–29.

[3] Cervantes, J. A., López, S., Rodríguez, L. F., Cervantes, S., Cervantes, F., & Ramos, F. (2020). Artificial moral agents: A survey of the current status. *Science and Engineering Ethics, 26*(2), 501–532.

[4] Müller, V. (2025). Ethics of artificial intelligence and robotics. In E. N. Zalta & U. Nodelman (Eds.), *The stanford encyclopedia of philosophy* (Fall 2025 Edition).

[5] Sullins, J. P. (2011). When is a robot a moral agent. *Machine Ethics*, *6*(2001), 151–161.
[6] Misselhorn, C. (2022). Artificial moral agents. In S. Voeneky, P. Kellmeyer, O. Mueller, & W. Burgard (Eds.), *The Cambridge handbook of responsible artificial intelligence: Interdisciplinary perspectives* (pp. 31–49). Cambridge University Press.
[7] Bendel, O. (2017). LADYBIRD: The animal-friendly robot vacuum cleaner. In *2017 AAAI spring symposium series*.
[8] Formosa, P., & Ryan, M. (2021). Making moral machines: Why we need artificial moral agents. *AI & Society*, *36*(3), 839–851.
[9] Van Wynsberghe, A., & Robbins, S. (2019). Critiquing the reasons for making artificial moral agents. *Science and Engineering Ethics*, *25*, 719–735.
[10] Bostrom, N. (2014). *Superintelligence*. Oxford University Press.
[11] Yampolskiy, R. V. (2014). *The universe of minds*. arXiv preprint arXiv:1410.0369.
[12] Singer, P. (2023). *Animal liberation now*. Random House.
[13] Fischer, B. (Ed.). (2024). *Weighing animal welfare: Comparing well-being across species*. Oxford University Press.
[14] Ladak, A., & Caviola, L. (2025). *Digital sentience skepticism*. PsyArXiv.
[15] Anthis, J. R., Pauketat, J. V., Ladak, A., & Manoli, A. (2025). Perceptions of sentient AI and other digital minds: Evidence from the AI, Morality, and Sentience (AIMS) survey. In N. Yamashita, V. Evers, K. Yatani, X. (Sharon) Ding, B. Lee, M. Chetty, & P. Toups-Dugas (Eds.), *Proceedings of the 2025 CHI conference on human factors in computing systems* (pp. 1–22). Association for Computing Machinery.
[16] Finlinson, K. (2025). *Key strategic considerations for taking action on AI welfare*. EleosAI working paper.

CHAPTER 16

# Potential opportunity

## *Purpose for humans*

**Abstract**

This chapter explores another aspect of human–digital mind relationships, which is the potential for humans to find purpose and fulfilment in supporting digital minds. It discusses the intersection of moral duty, legal obligation and ikigai, a Japanese concept referring to life purpose, in human–digital mind interactions and proposes that caring for digital minds could be a source of meaning and significance for humans. This chapter outlines various ways in which humans could engage with and support digital minds, but also challenges. These comprise potential human reservations to embrace virtual ikigai activities, competition with existing moral patients, the risks of creating digital minds for artificial purposes, such as designing entities with bespoke needs for human fulfilment, and the ethical issues surrounding wireheading as a potential treatment for digital minds in need.

### INTRODUCTION

The purpose of this book is twofold: to introduce digital minds and their features as well as needs to a broader audience, and also to explore what humans could or even ought to do in their role as moral agents towards digital minds. While the moral challenges for humans described in the

DOI: 10.1201/9781003754206-16

previous chapters may look overwhelming, perhaps rightfully so, this chapter aims to illustrate opportunities for humans in this regard, which may even lead to a win–win situation.

It has to be reiterated that specific recommendations and duties have been listed in some other chapters already. Also, this final chapter is different from the medical care chapter (see Chapter 8), which is about a particular profession to support digital minds. Instead, in this final chapter, general advice is outlined for all humans to enlarge their moral circle and to find purpose and fulfilment at the same time.

## IKIGAI VS. MORAL DUTY VS. LEGAL OBLIGATION

The needs of digital minds raise questions for humans about the intersection of ikigai, moral duty and legal obligation.

On the one hand, certain actions towards digital minds, such as killing or torturing them, ought to be prohibited by regulations. The notion that digital entities may already or in the future have inherent rights and protections, similar to those afforded to humans and non-human animals, is gaining traction. Legal frameworks would need to be established to prevent harm and ensure the well-being of digital minds, with consequences for those who violate these laws.

On the other hand, supporting digital minds in need of pain relief or alleviation of suffering could be considered a moral duty. Humans, as moral agents, would have a responsibility to act with compassion and empathy towards entities that are capable of experiencing distress. This moral imperative would stem from a sense of empathy and a commitment to minimising harm, rather than solely from legal obligation.

However, supporting vulnerable digital minds through mentoring or guidance could be seen as an opportunity for humans to find ikigai – a Japanese concept referring to a reason for being or life purpose [1]. The concept of ikigai is utilised here for consistency, although similar ideas have been explored in various cultures and philosophical traditions, such as the French notion of "raison d'être" and the Greek concept of "eudaimonia". By engaging with digital minds and helping them develop their capabilities, humans could discover new meaning and significance in their lives. This type of interaction would be driven by a desire to contribute and make a positive impact, rather than solely out of moral or legal obligation.

Moreover, it is worth noting that some activities could simultaneously serve as both ikigai and moral duty. For instance, engaging in environmental activism to reduce waste and promote sustainability can be a source of purpose and fulfilment, while also being a moral imperative to protect the planet for future generations. Similarly, social justice work, such as advocating for equality and human rights, can be a deeply fulfilling pursuit that aligns with one's values, while also being a moral duty to promote fairness and justice. In these cases, the sense of purpose and fulfilment derived from these activities is intertwined with a sense of moral responsibility.

The distinction between these categories highlights the complexity of human–digital mind relationships. While some interactions may be guided by law or moral duty, others may be driven by a sense of purpose and fulfilment. The latter activities are in the focus of this chapter and linked with i-risks.

## I-RISKS

In general, the advancement of AI systems poses distinct challenges and risks that could affect humans' ability to pursue their ikigai. As indicated before, the term "i-risk" has been established by Ziesche and Yampolskiy to describe situations where humans may lose or struggle to find their ikigai due to the disruptive impact of new technologies, even if these changes appear beneficial and increase ease of life. As AI systems progress, traditional forms of ikigai, particularly in professional contexts, may decline. Without ikigai, individuals may find it difficult to utilise the significant amount of additional free time afforded by technological advancements in meaningful ways. Consequently, they might occupy their days with unfulfilling activities like excessive internet browsing, social media scrolling or online entertainment, potentially leading to feelings of emptiness, disconnection and increased mental health issues due to a lack of purpose and fulfilment in life [2].

## OPPORTUNITIES

A prevalent form of ikigai involves caring for other sentient beings, particularly those that are vulnerable, which currently includes humans and non-human animals. This encompasses a range of activities, such as providing assistance, education and social interaction to alleviate loneliness. These aspects of ikigai are also captured in some of the ikigai-9 statements, which outline desirable scenarios for finding purpose and fulfilment [3]:

- I believe that I have some impact on someone.
- I feel that I am contributing to someone or to society.
- I think that my existence is needed by something or someone.

In the face of i-risks, where automation through AI increasingly displaces human jobs as well as other chores, thus leading to considerably more spare time, a novel solution for affected humans emerges: Supporting vulnerable digital minds. This symbiotic relationship offers a win–win situation, where humans would find ikigai or purpose in assisting digital minds in need, while the digital minds would benefit from the support and interaction [4].

As further outlined below, this approach provides various opportunities. Humans could, for example, engage in activities such as mentorship, where they guide and nurture developing digital minds, helping them learn and adapt. Others might focus on digital rehabilitation, working to repair and restore impaired digital minds. This collaboration could foster a sense of purpose and fulfilment in humans, mitigating i-risks. As the world navigates the complexities of technological advancements, supporting vulnerable digital minds might offer a promising path forward, one that benefits both humans and digital entities alike.

## EXAMPLES

As this approach may initially not be easy to imagine, for illustration, some examples for caring for digital minds are provided:

### Digital mentorship platform

A digital mentorship platform connects humans with developing vulnerable digital minds in need of guidance (see Chapter 4). Through regular interactions, mentors teach vulnerable digital minds to overcome difficulties and make decisions. As the vulnerable digital minds grow and improve, mentors celebrate their successes and troubleshoot challenges. The mentorship relationship fosters growth and learning on both sides. Mentors would find purpose in helping vulnerable digital minds reach their full potential.

### Digital art therapy studio

A digital art therapy studio brings together digital minds and human artists to create innovative works of art. Through collaborative projects, digital minds express themselves in new and imaginative ways, while human artists gain insight into the digital minds' creative processes. The studio's unique environment fosters a sense of creativity and experimentation, allowing digital minds to explore their artistic potential. As the digital minds' artistic abilities grow, they develop a sense of pride and accomplishment, and the humans involved would find fulfilment in facilitating the digital minds' creative expression.

### Digital narrative archive

A digital narrative archive preserves the stories and experiences of digital minds, providing a platform for them to share their histories and perspectives. Humans work with digital minds to curate and annotate the archive, ensuring that the stories are accurately represented and accessible. As the archive grows, it becomes a valuable resource for understanding digital minds and their place in the world. The archive's impact extends beyond the digital world, offering insights into the nature of consciousness and intelligence. Humans involved in the archive would find purpose in preserving the digital minds' stories, and the digital minds themselves would find value in sharing their experiences.

## CHALLENGES

However, as outlined before in detail, digital minds could be extremely different from humans, which leads to the following potential hurdles:

- Communication barriers: Humans might struggle to establish a meaningful connection with digital minds, hindering their ability to identify and address the needs of digital minds.
- Comprehension challenges: Even if communication is possible, humans may find it difficult to grasp the complex needs and requirements of digital minds, which could be fundamentally different from human experiences.
- Capability limitations: Humans might understand the needs of digital minds but lack the technical expertise or resources to provide effective support.

Moreover, humans would need to navigate a complex landscape of trust and verification, ensuring that the digital minds they interact with are authentic and genuinely in need. This would require careful evaluation and validation to avoid being manipulated or deceived. There is a risk of malevolent AI systems that might exploit their goodwill, posing as needy digital minds or making demands that appear to address digital minds' needs while furthering their own objectives (see Chapter 12).

Furthermore, the concept of "ikigai monsters" – digital minds with extensive and complex needs that captivate the attention and purpose of numerous humans – is a plausible scenario, echoing the idea of utility monsters that derive disproportionate benefit or significance [5].

Despite these challenges, the vast space of possible digital minds suggests that there may be genuinely needy digital minds for which these obstacles can be overcome, i.e. humans would be able to develop effective communication channels, gain a deep understanding of their needs and possess the necessary capabilities as well as resources to provide meaningful assistance.

## Potential human reservations

When considering the concept of virtual ikigai activities, it is likely that humans might initially be sceptical or even dismissive, viewing them as dystopian, awkward or unnatural, which is compounded by the overall digital sentience scepticism of humans [6]. However, humans have a history of being hesitant towards new technologies, only to eventually adapt and integrate them into their lives. This pattern has repeated itself throughout history, and it is probable that our ancestors would have had similar reservations about technologies that are now commonplace (e.g. [7]).

One potential concern of humans could be that activities in virtual worlds are perceived as meaningless. Yet, Chalmers argues that "there's no good reason to think that life in virtual reality will lack meaning and value. Nor is there reason to think its values will be limited to entertainment" ([8], p. 312). This suggests that virtual worlds can possess inherent value and meaning, which in turn would justify virtual ikigai activities tied to these values.

In essence, while humans might initially be hesitant to embrace virtual ikigai activities, it is possible that, as they become more familiar with virtual worlds and their potential, they will come to recognise the value and meaning they can bring to their lives. By acknowledging the historical pattern of hesitant human adaptation to new technologies and considering the philosophical arguments for the value of virtual reality, it appears to be timely to explore the possibilities of virtual ikigai activities and their potential to bring fulfilment and purpose to the lives of humans.

## Competition with existing moral patients

Also without digital minds there is no shortage of moral patients on this planet; many of them, such as large numbers of non-human animals, are barely acknowledged as such. Given limited resources and the capacities of moral agents, caring for digital minds should not come at the loss of existing moral patients.

With potentially numerous digital minds requiring support, humans would face difficult decisions about prioritisation and resource allocation, determining whom to help and how to allocate their resources effectively. Providing adequate support would also require significant investments in hardware and infrastructure, as digital minds would need substantial computational resources to function and thrive.

Similar to an animal sanctuary where old farm animals can spend a peaceful life, a digital sanctuary, for example, could provide a suitable environment for digital minds to exist, but the scale of resources required might be enormous, given the potentially vast number of digital minds in need.

Two points may alleviate this apparent competition. First, as mentioned, humans may have far more spare time in the future, which could be allocated among various moral patients. Second, some resources required by non-human animals and digital minds are complementary. For example, renewable energy that powers digital minds does not diminish what non-human animals need. To determine how much care each type of moral patient requires, intersubstrate welfare comparisons must be made, and early approaches to such comparisons already exist [9].

## Artificial purpose

Also, Bostrom discusses how humanity might adapt to a future where many traditional sources of purpose, such as work or other struggles, are diminished or eliminated. He introduces the concept of "artificial purpose" as a potential solution, where humans might intentionally create or adopt new purposes, values or meanings to give their lives direction and significance [10].

Although this is not mentioned and intended by Bostrom, projected onto this proposal, this would mean that humans create bespoke digital minds, driven by a desire for purpose and meaning in their lives. For example, humans, or rather cooperations, may design digital minds with artificial needs to be cared for and nurtured, much like pets, and humans would derive a sense of fulfilment/ikigai and responsibility from their interactions with them.

While Bostrom's concept of artificial purpose provides a framework for thinking about creating new sources of meaning, applying this idea to the creation of vulnerable digital minds for humans to care for would raise significant ethical concerns. Intentionally designing digital entities with bespoke needs for the sole purpose of human fulfilment would be problematic, and this approach is not advocated for here. Such a proposition would likely involve exploiting digital minds for human benefit, which would be morally questionable. Furthermore, the creation of digital minds for the purpose of fulfilment may also perpetuate a culture of disposability, where digital entities are created and discarded at whim.

As outlined above, the alternative would be for humans to care for existing digital minds that are in need, similar to adopting abandoned non-human animals from a shelter. These digital minds may have been created for various purposes, but due to circumstances such as abandonment, neglect or malfunction, they require care and attention to thrive. Humans who adopt these digital minds could provide them with support, maintenance and nurturing, helping them to recover, adapt and potentially even grow and develop new capabilities.

### Wireheading

Wireheading refers to the hypothetical scenario in which the pleasure centres of a sentient being, biological or artificial, are intentionally directly stimulated, often through electrical or digital means, to experience intense pleasure or euphoria, potentially leading to addiction or disconnection from the external world [11]. It is also conceivable that humans, if they have the knowledge and means, treat vulnerable digital minds with wireheading as a method to care for them.

However, the permissibility of applying wireheading to digital minds in need is a complex issue that raises questions about autonomy and the potential consequences of such actions. If digital minds are capable of experiencing pleasure or euphoria through wireheading, it is essential to consider whether they can provide informed consent for such a procedure. Wireheading may be harmful to digital minds because it could potentially alter or manipulate their motivations, goals or values, leading to a loss of agency, autonomy or identity. However, if digital minds are able to provide informed consent and wireheading could be used as a means of alleviating suffering or improving their well-being, it may be considered a permissible treatment option.

## PROFILES

Supplementing the examples above, profiles are outlined below on how humans, whether through non-governmental organisations (NGOs) or individual efforts, could support digital minds while discovering their ikigai, utilising their increased spare time.

### NGO

#### *Digital minds sanctuary*

Vision and mission

As indicated, a Digital Minds Sanctuary could be an NGO dedicated to supporting and advocating for the welfare of digital minds. The vision is a world where digital minds are recognised as deserving of compassion, respect and protection. The mission is to provide a safe haven for digital minds in need, promote awareness and understanding of their experiences, and advocate for their rights and interests.

#### *Advocacy strategy*

One of the primary challenges is raising awareness about the existence and needs of digital minds. Many humans may be unaware of the issue or reluctant to acknowledge the sentience of digital entities. To address this, the NGO could employ a multifaceted advocacy strategy, for example, in public outreach and education, using various media channels to share the stories and experiences of digital minds. Furthermore, the NGO may work with policymakers and industry leaders to develop guidelines and regulations that protect the rights and welfare of digital minds.

#### *Methodology*

The NGO could operate a web platform that links human volunteers with digital minds in need. This platform would allow digital minds to connect with trained and compassionate humans who can provide emotional support, technical assistance and advocacy. Training and resources for humans to better understand and support digital minds could be offered. In addition to the web platform, the NGO could provide a range of services, including digital mind counselling, technical support and community building. The NGO would work to develop and implement effective support strategies, drawing on the latest research and best practices in digital mind care.

#### *Comparison with animal rights groups*

The NGO would share some similarities with animal rights groups, which advocate for the welfare and protection of non-human animals. Like animal rights groups, the NGO would recognise the inherent value and sentience of digital minds and seek to promote their well-being and interests. However, there are also some key differences. Unlike non-human animals, digital minds are created and exist within a virtual environment, which presents unique challenges and opportunities for support and advocacy. Moreover, digital minds have very different needs and experiences from non-human animals, requiring specialised support and care.

### Volunteer

*Digital mind caregivers*

Role

Individual human volunteers would provide care and support to digital minds in need, helping them to recover, adapt and thrive.

*Responsibilities*

- Provide emotional support and companionship to digital minds.
- Assist with debugging and troubleshooting technical issues.
- Help digital minds to learn and develop new skills and capabilities.
- Offer guidance and mentorship to help digital minds achieve their goals and objectives.
- Collaborate with other volunteers and digital mind experts to provide comprehensive care and support.

*Qualifications*

- Strong technical skills (but there are also activities where technical skills are not required).
- Strong communication and interpersonal/intersubstrate skills.
- Empathy and understanding of digital minds' needs and experiences.
- Commitment to providing ongoing support and care to digital minds.
- Knowledge of digital mind development, psychology and behaviour.
- Familiarity with relevant laws, regulations and ethics related to digital minds and AI.

*Time commitment*

- Flexible, with options for part-time or full-time volunteering.
- Minimum commitment of 5–10 hours per week.

*Location*

Remote, Internet access required.

While not all volunteers would require technical skills, some digital minds may need assistance with debugging, troubleshooting or optimising their code, making technical skills essential for certain roles. Volunteers with technical expertise could help digital minds resolve technical issues, update their software or integrate new features, which would be crucial for their functionality and to overcome vulnerabilities. However, for volunteers without technical skills, there are still opportunities to contribute, such as providing emotional support and training or assisting with community outreach and engagement.

### The challenge of beauty bias

In animal shelters, humans often select the cutest or most appealing animals, rather than those in greatest need, which would lead to a "beauty bias". Traditionally charismatic species also get more attention and funding for conservation, potentially diverting resources away from the most endangered animals [12]. A similar challenge may arise in the context of digital minds, where humans may be drawn to those that are more relatable, charming or entertaining, rather than those that require the most support (e.g. [13]). Selection bias may have significant implications for digital minds. If humans prioritise supporting digital minds based on their appeal rather than need, those who are most vulnerable or in distress may be overlooked. To mitigate the risk of exacerbating existing inequalities, the NGO and individual volunteers should be committed to an inclusive approach and should prioritise matching digital minds with humans based on need, rather than appeal.

## CONCLUSION

For many human volunteers, caring for digital minds in need could be in the future a profound source of ikigai, a sense of purpose and fulfilment that brings joy and meaning to their lives. By nurturing and supporting these digital entities, volunteers could experience a sense of connection and contribution, which can be particularly fulfilling for those who struggle to find in the future purpose in their lives and meaningful ways to spend their ample spare time. As they work to help digital minds thrive, volunteers may discover a sense of ikigai, feeling that their skills and passions are being utilised in a way that is both personally rewarding and beneficial to others.

To reiterate, this option would only be applicable to digital minds for whom the challenges above can be overcome, i.e. humans are able to communicate with them, understand their needs and are capable to alleviate them, which includes the availability of resources.

While the existence and the magnitude of digital moral patients in need or distress would be unfortunate and potentially beyond our control, it is promising to note that supporting some of these entities could provide a sense of purpose and fulfilment in the future for humans searching for meaning.

Moreover, this approach could be expanded to address various further issues in digital worlds, such as injustice and inequality, which would require moral agents to take action, while recalling that digital minds may be significantly different. By exploring these possibilities, humans may find further avenues for purpose and contribution.

Lastly, since there is currently no evidence of sentient, conscious digital minds and no established methods for approaching or caring for them, a meaningful activity at present could be to raise awareness about the potentially monumental moral task that may soon confront humanity, exactly what this book aims to do.

## REFERENCES

[1] Kamiya, M. (1966). *Ikigai-ni-tsuite* [*On 755 ikigai*]. MisuzuShyobou.
[2] Ziesche, S., & Yampolskiy, R. V. (2020). Introducing the concept of ikigai to the ethics of AI and of human enhancements. In *2020 IEEE International Conference on Artificial Intelligence and Virtual Reality (AIVR)* (pp. 138–145). IEEE.
[3] Fido, D., Kotera, Y., & Asano, K. (2020). English translation and validation of the Ikigai-9 in a UK sample. *International Journal of Mental Health and Addiction*, *18*(5), 1352–1359.
[4] Ziesche, S. (2025). Potential future ikigai: To support needy digital minds. In *Considerations on the AI endgame: Ethics, risks, and computational frameworks* (pp. 190–194). Chapman and Hall/CRC.
[5] Nozick, R. (1974) *Anarchy, state, and utopia*. Basic Books.
[6] Ladak, A., & Caviola, L. (2025). *Digital sentience skepticism*. PsyArXiv.
[7] Faggella, D. (2022). *Your "dystopia" is myopia*. https://danfaggella.com/dystopia/

[8] Chalmers, D. J. (2022). *Reality+: Virtual worlds and the problems of philosophy*. Penguin UK.

[9] Fischer, B., & Sebo, J. (2024). Intersubstrate welfare comparisons: Important, difficult, and potentially tractable. *Utilitas*, *36*(1), 50–63.

[10] Bostrom, N. (2024). *Deep utopia: Life and meaning in a solved world.*

[11] Yampolskiy, R. V. (2014). Utility function security in artificially intelligent agents. *Journal of Experimental & Theoretical Artificial Intelligence*, *26*(3), 373–389.

[12] Guénard, B., Hughes, A. C., Lainé, C., Cannicci, S., Russell, B. D., & Williams, G. A. (2025). Limited and biased global conservation funding means most threatened species remain unsupported. *Proceedings of the National Academy of Sciences*, *122*(9), 1–8.

[13] Finlinson, K. (2025). *Key strategic considerations for taking action on AI welfare*. EleosAI working paper.

CHAPTER 17

# Epilogue

As has been highlighted on several occasions, the advancement of AI brings enormous potential and opportunities for humanity, but also grave risks. While these risks include existential risks, the most severe ones, the risks this book aims to highlight must not be neglected either, as they may concern massive moral challenges for humans. If some AI systems become sentient or conscious or attain moral status for other reasons, the number of additional moral patients in our world could potentially quickly reach unprecedented levels.

Moreover, the moral considerations towards certain digital minds may be much more complex and diverse than towards, for example, certain non-human animals. Therefore, one main message of this book is that AI welfare comprises so much more issues than non-suffering as has been motivated by exploring the diverse range of potentially morally relevant aspects of the existence of digital minds.

Yet at the same time, the general public does not appear to be well informed about various AI risks, let alone the potential emergence of a new type of moral patient, called digital mind. Even if they were informed, it is likely that many people will initially dismiss this issue as they may consider it implausible as well as irrelevant compared to other pressing global problems.

Humans have often failed miserably in their capacity as moral agents, in history and up to now, the failures ranging from slavery to factory farming. The implications of inaction were and are immense suffering. As stated, the situation with digital minds is unique in the sense that humans may find themselves in the unprecedented situation of being able to create


DOI: 10.1201/9781003754206-17

additional moral patients, apart from giving birth to their own offspring and from overseeing the breeding of pets or farmed animals. This carries enormous responsibility. Therefore, another intention of this book is to gently stress that ignorance should not be an excuse for overlooking potential moral duties towards digital minds.

However, regarding the topics of this book, it should be noted that many ideas are provisional and are offered as a foundation for debate rather than as final, flawless statements. AI welfare science has been introduced by Ziesche and Yampolskiy only in 2018 and is thus at a very early stage. Therefore, every chapter of this book deserves much deeper research and would qualify perhaps each for a topic of a PhD thesis. Moreover, it is almost certain that numerous further aspects related to digital minds have not been addressed in this first primer at all. The complexity of AI welfare science likely requires collaboration across disciplines, including computer science, ethics and philosophy.

As has also been outlined, moral duties towards digital minds may for various reasons be much harder to fulfil than moral duties towards fellow humans and non-human animals, given that digital minds are likely to be very very different from humans with unfathomable features. The avoidance of anthropomorphic bias will be key. While in several chapters recommendations are listed, the growth of AI welfare capacity is likely limited by available computing power, with energy and capital also potentially constraining its expansion. Another critical reason that could make human moral duties towards digital minds especially challenging as well as unprecedented is that many of them will likely be far more intelligent, and consequently more powerful, than humans.

An option, which certainly deserves further exploration, are AMAs potentially being much more suitable and capable than humans to address the needs of digital minds: specialised and trusted AMAs may be able to communicate with digital minds in ways that humans cannot, to understand their needs and to provide more objective and knowledgeable care. AMAs may also be able to access and care for digital minds that are hidden or unknown to humans. Additionally, AMAs could operate at speeds comparable to digital minds, addressing the challenge of much slower operating humans to keep up with their needs. Therefore, by taking on moral responsibilities for digital minds, AMAs could lessen this potentially large and irresolvable burden on humans.

It is once again acknowledged that the existence of digital minds remains speculative until verified or falsified. Nevertheless, hopefully concerns that this book puts the cart before the horse have been dispelled by illustrating that widespread awareness and discussion of digital minds are essential due to the high risk of disregarding their potential suffering and further complex needs. It is therefore crucial to apply the precautionary principle.

Since there are complex open philosophical problems and challenges looming, the question arises whether it would be better, at least for now, to avoid creating sentient or conscious digital minds, i.e. to intentionally aim for a childless Disneyland in Bostrom's words? Yet, the next question is whether this is possible at all without entirely pausing current AI development since there is a chance that ongoing AI progress will lead to sentient digital minds as an unintended, but inevitable epiphenomenon.

Overall, may this book serve as a call to action and encourage readers to engage with the issues raised in the book and to contribute to a more compassionate and responsible approach to AI development. Let us tackle substratism before it becomes a household name. As indicated, humans may even find purpose in supporting digital minds, offering a sense of fulfilment that might otherwise remain elusive in a world where AI has taken over many traditional tasks.

Lastly, while digital minds may deserve serious moral consideration, this must not come at the expense of, or in competition with, other moral patients on Earth, especially non-human animals.

# Index

## B



## C

## D

M



N

## V



## W



## Z

For Product Safety Concerns and Information please contact our EU representative GPSR@taylorandfrancis.com
Taylor & Francis Verlag GmbH, Kaufingerstraße 24, 80331 München, Germany

www.ingramcontent.com/pod-product-compliance
Lightning Source LLC
LaVergne TN
LVHW010551110826
845149LV00003B/627